AIRO Springer Series

Volume 13

The **AIRO Springer Series** focuses on the relevance of operations research (OR) in the scientific world and in real life applications.

The series publishes peer-reviewed only works, such as contributed volumes, lectures notes, and monographs in English language resulting from workshops, conferences, courses, schools, seminars, and research activities carried out by AIRO, Associazione Italiana di Ricerca Operativa – Optimization and Decision Sciences: http://www.airo.org/index.php/it/.

The books in the series will discuss recent results and analyze new trends focusing on the following areas: Optimization and Operation Research, including Continuous, Discrete and Network Optimization, and related industrial and territorial applications. Interdisciplinary contributions, showing a fruitful collaboration of scientists with researchers from other fields to address complex applications, are welcome.

The series is aimed at providing useful reference material to students, academic and industrial researchers at an international level.

Should an author wish to submit a manuscript, please note that this can be done by directly contacting the series Editorial Board, which is in charge of the peer-review process.

THE SERIES IS INDEXED IN SCOPUS

Andreas Brieden · Stefan Pickl · Markus Siegle
Editors

Graphs and Combinatorial Optimization: from Theory to Applications

CTW 2023, Garmisch-Partenkirchen, Germany, June 20–22

AIRO
ASSOCIAZIONE ITALIANA DI RICERCA OPERATIVA
OPTIMIZATION AND DECISION SCIENCE

Springer

Editors
Andreas Brieden
Fakultät für Wirtschafts- und
Organisationswissenschaften
Universität der Bundeswehr München
Munich, Germany

Stefan Pickl
Fakultät für Informatik
Universität der Bundeswehr München
Munich, Germany

Markus Siegle
Fakultät für Informatik
Universität der Bundeswehr München
Munich, Germany

ISSN 2523-7047 ISSN 2523-7055 (electronic)
AIRO Springer Series
ISBN 978-3-031-46825-4 ISBN 978-3-031-46826-1 (eBook)
https://doi.org/10.1007/978-3-031-46826-1

This Springer imprint is published by the registered company Springer Nature Switzerland AG
The registered company address is: Gewerbestrasse 11, 6330 Cham, Switzerland

Paper in this product is recyclable.

CTW 2023 Program Committee

Ali Fuat Alkaya, Marmara U.
Andreas Brieden, U. Bundeswehr München
Brigitte Buchetmann, U. Bundeswehr München
Christoph Buchheim, TU Dortmund
Francesco Carrabs, U. Salerno
Alberto Ceselli, U. Milano
Roberto Cordone, U. Milano
Ekrem Duman, Ozyegin U.
Yuri Faenza, Columbia U.
Claudio Gentile, IASI-CNR Roma
Johann Hurink, U. Twente
Ola Jabali, Politecnico di Milano
Leo Liberti, CNRS and LIX Polytechnique Palaiseau
Bodo Manthey, U. Twente
Gaia Nicosia, U. Roma
Tre Tony Nixon, U. Lancaster
Andrea Pacifici, U. Roma Tor Vergata
Ulrich Pferschy, U. Graz
Stefan Pickl, U. Bundeswehr München
Bert Randerath, U. Köln
Giovanni Righini, U. Milano
Heiko Roeglin, U. Bonn
Oliver Schaudt, RTWH Aachen U.
Rainer Schrader, U. Köln
Markus Siegle, U. Bundeswehr München
Giuseppe Stecca, IASI-CNR Roma

Paolo Ventura, IASI-CNR Roma
Maria Teresa Vespucci, U. Bergamo
Angelika Wiegele, Alpen-Adria U. Klagenfurt
Maryna Zharikova, Kherson National Technical U.

CTW Steering Committee

Ali Fuat Alkaya, Marmara U.
Alberto Ceselli, U. Milano
Roberto Cordone, U. Milano
Ekrem Duman, Ozyegin U.
Ulrich Faigle, U. Köln
Claudio Gentile, IASI-CNR Roma
Johann Hurink, U. Twente
Leo Liberti, CNRS and LIX Polytechnique Palaiseau
Bodo Manthey, U. Twente
Gaia Nicosia, U. Roma
Tre Andrea Pacifici, U. Roma Tor Vergata
Stefan Pickl, U. Bundeswehr München
Bert Randerath, U. Köln
Giovanni Righini, U. Milano
Heiko Roeglin, U. Bonn
Oliver Schaudt, RWTH Aachen
Rainer Schrader, U. Köln
Rudiger Schultz, U. Duisburg-Essen
Frank Vallentin, U. Köln

CTW23 Organizing Committee

Dr. Andrea Ferstl
Tino Krug
Viola Schad
Ulrike Stein
Silvia Wagner

Program Committee Chairs

Andreas Brieden
Stefan Pickl
Markus Siegle

Preface

The Cologne-Twente Workshop (CTW) on Graphs and Combinatorial Optimization is an established workshop series initiated by Ulrich Faigle in 2001, which was the time he moved from Twente University to the University of Cologne. After several CTW editions in Twente and Cologne, the workshop was also held in different locations in Italy, France, Germany, the Netherlands, and Türkiye. Having been initially set up by discrete applied mathematicians, CTW still follows the mathematical tradition.

In this CTW2023 edition, we implemented again two submission tracks: standard papers of at most 12 pages and traditional CTW extended abstracts of at most 4 pages. This volume collects the standard papers that were accepted by CTW2023. The papers underwent a peer-review process performed by a Program Committee consisting of 30 members and 19 CTW steering committee members. PC members came from Austria, France, Germany, Italy, The Netherlands, Türkiye, Ukraine, UK, and the USA. We received 33 submissions of which 15 (45.5%) were accepted for publication in this volume. The chapters of this volume present works on graph theory, discrete mathematics, combinatorial optimization, and operations research methods, with particular emphasis on coloring, graph decomposition, connectivity, distance geometry, mixed-integer programming, machine learning, heuristics, metaheuristics, math-heuristics, and exact methods. Applications are related to logistics, production planning, and scheduling.

The scientific program of CTW2023 included presentations of the 15 standard papers in this volume, of 10 extended abstracts, and 4 plenary invited lectures. As usual for the CTW, extended abstracts were subject to a high acceptance level, allowing also papers containing preliminary results with a particular focus on work presented by young researchers. Those traditional CTW extended abstracts were published in an internal publication and on the conference's website at www.ctw2023.de.

We thank all PC members for their hard reviewing work performed to select the papers and to improve their quality.

Following the CTW tradition, a special issue of Discrete Applied Mathematics (DAM) journal dedicated to this workshop and its main topics of interest will be edited.

This CTW edition also featured invited plenary speakers. Four well-known researchers accepted our invitation: Prof. Peter Gritzmann (Technische Universität München) spoke about "Diagrams, clustering, and coresets, and their application to the representation of polycrystals", Prof. Janny Leung (University of Macau) gave insights into the complex nature of "Sports Scheduling", Prof. Anne Remke (Universität Münster) gave an overview of "Optimizing different flavours of nondeterminism in hybrid automata with random clocks", and Prof. Maximilian Moll (Universität der Bundeswehr München) chose the topic "Exploring Solutions to the Interdiction Problem: Network Optimization in Operations Research, Machine Learning and Quantum Computing".

We would like to thank the Associazione Italiana di Ricerca Operativa (AIRO) for hosting this volume in its AIRO Springer series. Also, last but not least, we would like to thank our organizing team for their dedicated work in making this workshop possible.

Munich, Germany
July 2023

Andreas Brieden
Stefan Pickl
Markus Siegle

Contents

The Algorithmic Complexity of the Paired Matching Problem

Ruben F. A. Verhaegh

Abstract We introduce a new matching problem originating from industry called the Paired Matching problem. The objective in the problem is to find a maximum matching of minimum cost in a bipartite graph. This is complicated by a non-trivial definition of cost, which is expressed based on a pairing of the vertices in one partite set. We prove that the problem is NP-complete even under further restrictions. We also study the parameterized complexity of the problem and give an exact algorithm for it using kernelization. In doing so, we show that the problem can be solved efficiently even on large inputs, as long as a given one of the partite sets is small.

1 Introduction

Background and motivation. The field of matching theory studies and explores matchings (sets of pairwise vertex-disjoint edges) and their properties, which are relevant for a plethora of applications. Some of the most fundamental matching problems, such as the Maximum Matching problem and the Minimum Weight Perfect Matching problem, have long been known to be solvable in polynomial time [4], while many other matching problems however have been proven to be NP-complete, including the Exact Weight Perfect Matching problem [10], the Induced Matching problem [11] and the Rainbow Matching problem [7].

We introduce and study a new matching problem. Its origins and relevance lie in pick-and-place machines that are used in electronics assembly to place electrical components on printed circuit boards. Recently, a company from the Eindhoven Brainport area designed an upgrade to one of their machines, such that it could pick and place two of these components at the same time. With this upgrade also comes a new optimization problem: which pairs of components should be handled simul-

This work is based on a Masters thesis [12].

R. F. A. Verhaegh (✉)
Eindhoven University of Technology, Eindhoven, The Netherlands
e-mail: r.f.a.verhaegh@tue.nl

A. Brieden et al. (eds.), *Graphs and Combinatorial Optimization: from Theory to Applications*, AIRO Springer Series 13,
https://doi.org/10.1007/978-3-031-46826-1_1

taneously to minimize the time spent? This optimization problem can be formulated as the following problem.

Minimum Paired Matching (MPM)

Given: A bipartite graph $G = (P + T, E)$ with "people vertices" P and "task vertices" T, where the $|P| = 2n$ people vertices are partitioned into ordered pairs $(a_1, b_1), (a_2, b_2), \ldots, (a_n, b_n)$. Also given are a constant $c \in \mathbb{R}_{\geq 0}$, and a function $f : T \times T \to \mathbb{R}_{\geq 0}$

Objective: Determine the minimum cost of a matching with cardinality $|T|$ in G or conclude that no such matching exists in G. Each people pair (a_i, b_i) contributes separately to the cost of a matching M and the total cost of the matching is the sum of all these contributions. The contribution of a pair (a_i, b_i) is as follows:

- If neither a_i nor b_i is matched to a task by M, their contribution to the cost is 0.
- If exactly one of a_i and b_i is matched to a task by M, their contribution to the cost is c.
- If a_i is matched to task $t_1 \in T$ and b_i is matched to task $t_2 \in T$, then their contribution to the cost is given by $f(t_1, t_2)$. Note the possible asymmetry in the definition of f, meaning that $f(t_1, t_2)$ might not equal $f(t_2, t_1)$.

We define the PAIRED MATCHING problem (PM) to be the corresponding decision problem in which an additional parameter $d \in \mathbb{R}_{\geq 0}$ is added to the input. Rather than finding the minimum cost of a matching with cardinality $|T|$ in the graph, the objective is then to determine whether such a matching exists with cost at most d.

For the practical application in electronics manufacturing, one may think of the tasks in this definition as electrical components and think of the people as machine parts to pick and place these components. Then, the cost to be minimized represents the total time spent to complete the assembly for a given choice of assigning electrical components to machine parts.

Despite the practically motivated origins of the problem, there are some deep links between PM and other better known matching problems, in particular to the EXACT MATCHING problem [10]. Although not part of this paper, the relation between PM and existing literature is explored further in the master thesis this work is based on [12].

Our contribution. In Sect. 2, we consider the PAIRED MATCHING problem from the perspective of classical complexity theory: we prove the decision problem PM to be NP-complete and make some notes on the implications this has on the (in)approximability of the optimization problem MPM.

In Sect. 3, we shift our focus to the parameterized complexity of PM. By exploiting a kernelization technique presented by Bodlaender, Jansen and Kratsch [1], we give an exact algorithm that solves a PM instance I with $|T|$ task vertices in time

$2^{\mathcal{O}(|T|^3)}|I|^{\mathcal{O}(1)}$. Hence this algorithm shows that large instances with a small number of task vertices can be solved efficiently.

Parameterized complexity of other matching problems. As briefly mentioned above, PM is closely related to the EXACT MATCHING problem. This problem itself has also been studied from the perspective of parameterized complexity theory. In [8] for example, parameterized algorithms for it were developed when considering the independence number or bipartite independence number of the graph to be the parameter. Another common choice of parameter for graph problems is, if applicable, the size of the solution whose existence needs to be determined. Since the objective in the EXACT MATCHING problem is to determine the existence of a specific type of *perfect* matching, this size is always polynomial in the input size, making the solution size an uninteresting parameter to consider for EM.

Many other matching problems however ask for matchings that are not necessarily perfect and problems like the RAINBOW MATCHING problem [6] and 3-DIMENSIONAL MATCHING problem [5] admit efficient parameterized algorithms when parameterized by the solution size. Contrarily, the INDUCED MATCHING problem is an example of a matching problem for which it has been proven that such algorithms are unlikely to exist [9].

Notation. Throughout the paper, we use standard graph notation. Any notation not defined here can be found in the book Parameterized Algorithms by Cygan et al. [3]. Although our results are presented in such a way that familiarity with the field of parameterized algorithms is not required, the book may also also be a good reference for readers unfamiliar with the field to provide additional background and motivation.

2 Classical Complexity

We start this section by proving PM to be NP-complete in Theorem 1. Afterwards, we discuss the implications of the theorem for the approximability for MPM.

Before proving the NP-hardness of PM, briefly note that the problem is contained in NP: the cost of any matching in a PM instance can be computed in polynomial time, so a solution to the problem may be verified in polynomial time as well. Now to prove that the problem is also NP-hard, we provide a polynomial time reduction from 3OCC-SAT, a variant of the well-known SAT problem which asks to determine the satisfiability of a Boolean formula in conjunctive normal form. The 3OCC-SAT problem poses an extra restriction on the input by requiring that every variable occurs at most three times.

3-Occurence Satisfiability (3OCC-SAT)

Given: A Boolean formula F in conjunctive normal form (CNF) on the variables $X = \{x_1, x_2, \dots, x_n\}$ such that every variable occurs at most three times
Objective: Determine whether X admits a truth assignment that satisfies F.

Although many variants of the SAT problem limit the clauses to contain no more than three literals, remark that clauses in 3OCC-SAT instances are allowed to be arbitrarily large. Now, given the NP-completeness of the SAT problem [2], it is not hard to convince ourselves of the NP-completeness of 3OCC-SAT as every CNF formula can be rewritten to an equivalent one in which no variables occur more than three times [13]. Hence, we can use this problem to prove Theorem 1.

Theorem 1 *Let $c_1, c_2 \in \mathbb{R}_{\geq 0}$ be any two constants such that $0 \leq c_1 < c_2$. Then any 3OCC-SAT instance can be reduced to an equivalent PM instance in which $c = c_1$, f only takes values $2 \cdot c_1$ and $2 \cdot c_2$ and $d = c_1 \cdot |T|$. Hence, PM is NP-complete, even when c and f are integer valued and bounded by a constant.*

Proof Let F be a 3OCC-SAT instance, where F is a formula with clauses $C_1, C_2, \dots, C_m$ using the variables $X = \{x_1, x_2, \dots, x_n\}$. Let $\ell_{i,1}, \ell_{i,2}, \dots, \ell_{i,|C_i|}$ be the occurrences of literals of the i-th clause. We reduce F to an equivalent PM-instance (G, c, f, d) and we start by constructing the graph $G = (P + T, E)$. First, the set T will be constructed as the union of two sets T_1 and T_2.

For every clause C_i we add a vertex v_i to T_1. For every occurrence $\ell_{i,j}$ of a literal we add two vertices $p^+_{i,j}$ and $p^-_{i,j}$ to P, representing a TRUE or FALSE assignment of the variable in $\ell_{i,j}$ respectively. These two vertices form a pair in P. If $\ell_{i,j}$ is a positive variable we connect $p^+_{i,j}$ and v_i. If $\ell_{i,j}$ is a negated variable we connect $p^-_{i,j}$ and v_i. Remark that this construction thus creates multiple people pairs for literals occurring multiple times.

By finding a matching in this graph which matches all vertices in T_1, we are determining an assignment of the variables in F which satisfies all its clauses. However, not every such matching in G necessarily corresponds to a valid truth assignment of the variables in X: we might try to set a variable to TRUE to satisfy one clause and at the same time set it to FALSE to satisfy another clause. Hence, we need to expand our graph to prevent situations like these. We can only encounter this problem for variables that occur both positively and negatively in the formula, so we expand our reduction based on the following two cases:

- Suppose x_h is a variable that occurs twice in F: once positively in the literal ℓ_{i_1,j_1} and once negated in the literal ℓ_{i_2,j_2}. We then add a vertex s_h to the set T_2 and connect $p^+_{i_1,j_1}$ and $p^-_{i_2,j_2}$ to s_h. See Fig. 1a for an example.
- Suppose x_h is a variable that occurs three times in F and suppose that it occurs positively in the literals ℓ_{i_1,j_1} and ℓ_{i_2,j_2} and that it occurs negated in the literal ℓ_{i_3,j_3}. We then add two vertices $s_{h,1}$ and $s_{h,2}$ to T_2. We connect both $p^+_{i_1,j_1}$ and $p^+_{i_3,j_3}$ to $s_{h,1}$ and we connect both $p^+_{i_2,j_2}$ and $p^-_{i_3,j_3}$ to $s_{h,2}$. See Fig. 1b for an example. For

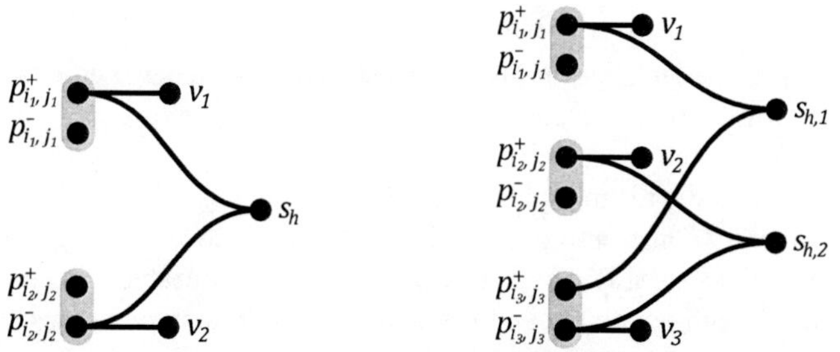

(a) The direct neighborhood of the vertices representing the literals that x_h occurs in if it occurs once positively and once negatively.

(b) The direct neighborhood of the vertices representing the literals that x_h occurs in if it occurs twice positively and once negatively.

Fig. 1 The structure created for variables that occur two or three times in F

variables that occur once positively and twice negated, we can simply swap the $+$ and $-$ signs in this step.

For variables that occur only in positive form or only in negated form, we skip this step as there is only one assignment for such a variable that allows it to satisfy clauses. The problematic situation explained above, where it is simultaneously assigned TRUE to satisfy one clause and FALSE to satisfy another, is therefore not applicable to these variables.

By taking $T = T_1 \cup T_2$ we have now finalized the construction of $G = (P + T, E)$. Now let c_1 and c_2 be any two given constants with $0 \leq c_1 < c_2$. We let c and f depend on these two constants by taking $c = c_1$ and taking f to be given by:

$$f(t_1, t_2) = \begin{cases} 2c_1 & \text{if } t_1, t_2 \in T_2 \\ 2c_2 & \text{otherwise.} \end{cases}$$

Finally, we take $d = c_1 \cdot |T|$, to complete our construction of the PM instance (G, c, f, d). Observe that this reduction can be done in polynomial time. To prove that it is also correct we show that F is a YES-instance if and only if (G, c, f, d) is a YES-instance.

($\Rightarrow$) Suppose that F is a YES-instance for 3OCC-SAT. Then there is a truth assignment $\mathcal{T} : X \to \{\text{TRUE}, \text{FALSE}\}$ of the variables in X that satisfies F. In particular, it satisfies each clause individually. We will use this assignment to construct a matching M that covers every vertex in T and which has a cost of d. We construct it as the union of two matchings M_1 and M_2. M_1 is used to cover the vertices in T_1 and can be seen as actually encoding the truth assignment $\mathcal{T}$ into G. M_2 is used to cover the remaining vertices at a low enough cost. First we construct M_1.

Consider the clause C_i and consider a literal $\ell_{i,j}$ that satisfies it under assignment $\mathcal{T}$. If there are multiple literals satisfying the clause, we can pick any arbitrary one of them. If $\ell_{i,j}$ is a positive variable, we include $\{p^+_{i,j}, v_i\}$ in M_1. If $\ell_{i,j}$ is a negated variable, we include $\{p^-_{i,j}, v_i\}$ in M_1. By construction, whichever edge we add, exists in G. Doing this for every clause C_i ensures that every vertex in T_1 is covered by M_1. Furthermore, the vertices in T_1 do not have any shared neighbors so none of the edges of M_1 coincide, meaning it is indeed a matching.

We continue by constructing a matching M_2 which covers the vertices in T_2 and none of the vertices already covered by M_1. Consider a variable x_h that occurs more than once in F, both positively and negated. We distinguish two cases:

- Suppose x_h occurs twice in F: once positively in ℓ_{i_1,j_1} and once negatively in ℓ_{i_2,j_2}. If x_h is set to TRUE in $\mathcal{T}$, we know that ℓ_{i_2,j_2} cannot be used to satisfy C_{i_2}, which in turn means that we did not cover $p^-_{i_2,j_2}$ with M_1. Hence, we can add $\{p^-_{i_2,j_2}, s_h\}$ to M_2. By construction, this edge exists. Similarly, if x_h is instead set to FALSE in $\mathcal{T}$, we add $\{p^+_{i_1,j_1}, s_h\}$ to M_2. Again, this edge exists and does not coincide with an edge from M_1.
- Suppose x_h occurs three times in F. Assume for now that it occurs twice positively in ℓ_{i_1,j_1} and ℓ_{i_2,j_2} and once negated in ℓ_{i_3,j_3}. If x_h is set to TRUE in $\mathcal{T}$, we know that ℓ_{i_3,j_3} cannot be used to satisfy C_{i_3}, which in turn means that we did not cover $p^-_{i_3,j_3}$ with M_1. Because $p^+_{i_3,j_3}$ does not have any neighbors in T_1, this vertex was also not covered by M_1. Hence we can add the edges $\{p^+_{i_3,j_3}, s_{h,1}\}$ and $\{p^-_{i_3,j_3}, s_{h,2}\}$ to M_2. By construction, these edges exist.
  If x_h is instead set to FALSE in $\mathcal{T}$, we know that ℓ_{i_1,j_1} cannot be used to satisfy C_{i_1} and ℓ_{i_2,j_2} cannot be used to satisfy C_{i_2}. This in turn means that neither $p^+_{i_1,j_1}$ nor $p^+_{i_2,j_2}$ was covered by M_1. Hence, we can add the edges $\{p^+_{i_1,j_1}, s_{h,1}\}$ and $\{p^+_{i_2,j_2}, s_{h,2}\}$ to M_2. By construction these edges exist.
  If x_h were to occur once positively and twice negated in F, we can simply swap the $+$ and $-$ signs in this step.

If we do this for every variable that occurs both positively and negated, we ensure that every vertex in T_2 is covered by M_2. Moreover, we have done so without covering vertices already covered by M_1. Combining this with the fact that M_1 and M_2 are both matchings, we get that $M := M_1 \cup M_2$ is also a matching. M then covers all vertices in $T = T_1 \cup T_2$.

Finally, we determine the cost of M. Note that most pairs in P have only one edge matched by M, therefore each contributing $c = c_1$ to the total cost of M. In our construction, the only pairs in P of which both vertices are matched by M correspond to variables that occur three times. In fact, every such variable has only one literal $\ell_{i,j}$ whose corresponding vertex pair even has edges connected to both vertices at all. There are only two cases in which both these vertices are matched by M: if $\ell_{i,j}$ is a positive variable which is set to FALSE in the satisfying assignment $\mathcal{T}$ or if $\ell_{i,j}$ is a negated variable which is set to TRUE in $\mathcal{T}$. In these cases, the corresponding vertices $p^+_{i,j}$ and $p^-_{i,j}$ are both matched to a vertex in T_2, meaning that together they contribute $2c_1$ to the cost of M. So for every edge in M it can be said that it contributes c_1 to the cost of M, meaning that its cost is $c_1 \cdot |T| = d$. Hence, (G, c, f, d) is a YES-instance.

($\Leftarrow$) Suppose that (G, c, f, d) is a YES-instance. Then there exists a matching M in $G = (P + T, E)$ which has cost at most $d = c_1 \cdot |T|$ and covers all vertices in T. We partition the matching into $M = M_1 \cup M_2$ such that M_1 contains all the edges covering T_1 and M_2 contains all the edges covering T_2. We show that there exists an assignment $\mathcal{T}$ of the variables $x_1, \ldots, x_n$ such that F is satisfied.

For every edge $\{p^+_{i,j}, v_i\} \in M$ we assign the variable in the literal $\ell_{i,j}$ to TRUE and for every edge $\{p^-_{i,j}, v_i\} \in M$ we assign the variable in the literal $\ell_{i,j}$ to FALSE. Any possible remaining variables may receive an arbitrary assignment. Because M matches every vertex in T and in particular in T_1, every clause of F is satisfied by this assignment: if $\{p^+_{i,j}, v_i\} \in M$ (and therefore in G), then by construction C_i contains the positive literal $\ell_{i,j}$ meaning that it can be satisfied by setting the corresponding variable to TRUE. Likewise, if $\{p^-_{i,j}, v_i\} \in M$ (and therefore in G), then by construction C_i contains the negated literal $\ell_{i,j}$ meaning that it can be satisfied by setting the corresponding variable to FALSE. So this assignment indeed satisfies F. It remains to show that this is a valid assignment, i.e.: there is no variable which is set to TRUE and FALSE to satisfy multiple clauses.

This problem could of course only happen for a variable x_h which occurs at least once positively and at least once negated. We distinguish two cases:

- Suppose x_h occurs twice in F: once positively in ℓ_{i_1,j_1} and once negated in ℓ_{i_2,j_2}. This implies the presence of a vertex $s_h \in T$ which only has the vertices $p^+_{i_1,j_1}$ and $p^-_{i_2,j_2}$ as neighbors. Since s_h is by definition covered by M, we know that at least one of these two neighbors must be connected to x_h in M and because M is a matching, this particular vertex is not also used to satisfy its corresponding clause. Hence, at most one of the literals ℓ_{i_1,j_1} and ℓ_{i_2,j_2} is used to satisfy its corresponding clause meaning that x_h does not get assigned both TRUE and FALSE.
- Suppose x_h occurs three times in F. Assume for now that it occurs twice positively in ℓ_{i_1,j_1} and ℓ_{i_2,j_2} and once negated in ℓ_{i_3,j_3}. We will prove that if $\{p^-_{i_3,j_3}, v_{i_3}\} \in M$ (in which case we assign x_h to FALSE to satisfy clause C_{i_3}) it holds that $\{p^+_{i_1,j_1}, v_{i_1}\}, \{p^+_{i_2,j_2}, v_{i_2}\} \notin M$. This implies that x_h is not set to TRUE and FALSE simultaneously to satisfy multiple clauses.
  So suppose that $\{p^-_{i_3,j_3}, v_{i_3}\} \in M$. Because $s_{h,2}$ is by definition matched in M, but apparently not to $p^-_{i_3,j_3}$, it must instead be matched to its only other neighbor: $p^+_{i_2,j_2}$. Because $p^+_{i_2,j_2}$ is then already matched to a vertex other than v_{i_2} by M and M is a matching, we know that $\{p^+_{i_2,j_2}, v_{i_2}\} \notin M$.
  To show that also $\{p^+_{i_1,j_1}, v_{i_1}\} \notin M$, we first argue that $\{p^+_{i_3,j_3}, s_{h,1}\} \notin M$. If this edge were to be in M, then the pair $(p^+_{i_3,j_3}, p^-_{i_3,j_3})$ would contribute $2c_2$ to the cost of M by the definition of f. Since M has size $|T|$ and a cost of at most $c_1 \cdot |T|$, pairs in P cannot contribute more than c_1 to the cost per matched vertex in them without exceeding the cost of d. Hence, $\{p^+_{i_3,j_3}, s_{h,1}\}$ cannot be in M as it would make the pair $(p^+_{i_3,j_3}, p^-_{i_3,j_3})$ contribute $c_2 > c_1$ to the cost of M per matched vertex. So we establish that $\{p^+_{i_3,j_3}, s_{h,1}\} \notin M$. This implies that $\{p^+_{i_1,j_1}, s_{h,1}\} \in M$, because $s_{h,1}$

must be matched by M and $p^+_{i_1,j_1}$ is its only other neighbor, which in turn implies that $\{p^+_{i_1,j_1}, v_{i_1}\} \notin M$ because M is a matching.
This shows that x_h is not set to TRUE and FALSE simultaneously in our assignment. An analogous argument can be made in case x_h occurs once positively and twice negated in F.

This proves that our assignment is indeed valid and satisfies F, meaning that F is a YES-instance for 3OCC-SAT. □

The additional constraints posed by Theorem 1 under which PM is still NP-complete, also have implications for the approximability of MPM.

Corollary 1 *Unless P=NP, MPM cannot be approximated in polynomial time to within a multiplicative factor of the optimal solution.*

Proof By taking $c_1 = 0$ in Theorem 1, we obtain a restriction to the input of PM in which we always have $d = 0$. An algorithm that approximates MPM within some multiplicative factor of the optimal solution would return 0 if and only if the optimal solution is 0, so this approximation algorithm could also be used to solve any PM instance under this restriction. Since Theorem 1 states that this restricted version of PM is still NP-complete, such an approximation algorithm cannot run in polynomial time unless P = NP. □

Of course, any optimization problem in which the optimal solution is 0 can be solved exactly using an approximation algorithm. Approximating such problems is therefore just as hard as solving them exactly. However, even if we require c and f to be positive in MPM, thereby avoiding the trivial case from above where $d = 0$, we find the problem to be hard to approximate [12].

3 Solving Large Instances with Few Task Vertices

In this next section, we consider the parameterized complexity of PM, where our parameter of choice is $|T|$, the number of task vertices in an instance. We develop an algorithm for PM based on *kernelization*, a popular technique for developing parameterized algorithms. A complete definition and explanation of the concept can be found in the book by Cygan et al. [3], but the high level idea is to reduce an instance I of a problem to an equivalent instance I' of the same problem whose size depends only on the parameter associated with the original instance I. Here, we say that the two instances are equivalent when they can be answered with the same YES/NO answer.

When the reduction can be executed in time that is polynomial in the original input size $|I|$, we call it a kernelization. Remark that any follow-up algorithm that is run on the newly created, yet equivalent instance I' has a running time depending only on the parameter associated with I, rather than the size of I.

We first show that a kernelization exists for PM with respect to $|T|$ in Proposition 1 and explain in Corollary 2 how it can be used to solve large PM instances with few task vertices efficiently.

The kernelization is based on the observation that it does not matter to which people pair a task is assigned. It only matters which other task vertex gets connected to the same pair, if any. This means that we could remove all people pairs from an instance for which it holds that for any solution covering that pair, there also exists another solution of the same cost which does not cover that pair. A key ingredient to prove the correctness of the kernelization will be a formalization of this idea as given by the following result from Bodlaender, Jansen, and Kratsch [1, (Theorem 2)].

Lemma 1 *Let $B = (X + Y, E)$ be a bipartite graph and let $M \subseteq E$ be a maximum matching in B. Let $X_M \subseteq X$ be the set of vertices in X that are covered by M. Then for each $Y' \subseteq Y$, if there exists a matching M' in B that covers Y', then there exists a matching M'' in $G[X_M \cup Y]$ that covers Y'.*

We continue by using it to show the following result:

Proposition 1 *Any PM instance (G, c, f, d) with $G = (P + T, E)$ can be reduced in polynomial time to an equivalent PM instance (G', c, f, d) in which G' has $\mathcal{O}(|T|^2)$ vertices and $\mathcal{O}(|T|^3)$ edges.*

Proof Let $(G = (P + T, E), c, f, d)$ be a PM instance. We will construct a graph $G' = (P' + T', E')$ with $\mathcal{O}(|T|^2)$ vertices and $\mathcal{O}(|T|^3)$ edges such that (G, c, f, d) is a YES-instance if and only if (G', c, f, d) is a YES-instance. Figure 2 shows the kernelization for an example input.

Before constructing G', we first construct an auxiliary bipartite graph $B = (X + Y, E'')$ from G. For every pair (a_i, b_i) in P we add a vertex x_i to X. We construct Y as the union of two sets $Y = Y_1 \cup Y_2$. For every vertex $t_j \in T$ we add a vertex y_j to Y_1 and for every pair of distinct vertices $t_{j_1}, t_{j_2} \in T$ we add the vertices y_{j_1,j_2} and y_{j_2,j_1} to Y_2. We add an edge between every $x_i \in X$ and $y_j \in Y_1$ for which either $\{a_i, t_j\} \in E$ or $\{b_i, t_j\} \in E$. We also add an edge between every $x_i \in X$ and $y_{j_1,j_2} \in Y_2$ for which $\{a_i, t_{j_1}\}, \{b_i, t_{j_2}\} \in E$. This concludes the construction of B.

An interesting observation to make is that every matching M_G in G can be encoded in B as a matching M_B. Given any M_G, we could construct a corresponding M_B in which:

- The presence of an edge between some $x_i \in X$ and some $y_j \in Y_1$ would indicate that M_G leaves one vertex of (a_i, b_i) unmatched, while matching the other to $t_j \in T$.
- The presence of an edge between some $x_i \in X$ and some $y_{j_1,j_2} \in Y_2$ in M_B would indicate that a_i and b_i get matched to t_{j_1} and t_{j_2} respectively by M_G.

Now that we have constructed our auxiliary graph B, we continue by finding a maximum matching M in it and we use this matching to construct the graph $G' = (P' + T', E')$. To this end, let $X_M \subseteq X$ denote the set of vertices in X that are

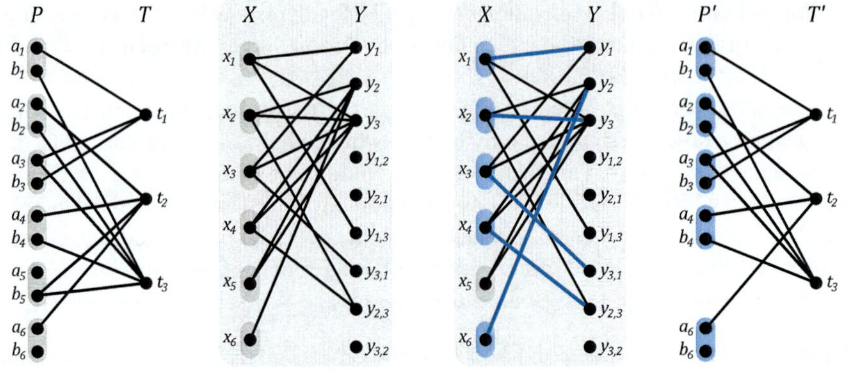

(a) *Input graph* G. Example of an unbalanced bipartite graph G.

(b) *Auxiliary graph* B. Step 1: construct the auxiliary graph B. Consider e.g. the vertices of (a_1, b_1). They are connected to t_1 and t_3 respectively in G, so x_1 is connected to y_1, y_3 and $y_{1,3}$ in B.

(c) *Auxiliary graph* B. Step 2: find a maximum matching M in B (the blue edges). The vertices in X_M are highlighted blue.

(d) *Output graph* G'. Step 3: remove people pairs from P corresponding to unmatched vertices in B to obtain G'.

Fig. 2 Kernelization applied to an example graph. Figure 2a shows an example graph G. Figures 2b and c depict the auxiliary graph B and respectively show its construction and a maximum matching in it. They are drawn on a gray background to avoid confusion between the three different graphs depicted (G, B and G'). Figure 2d shows the resulting graph G'

covered by M. Then G' is obtained from G by removing all people pairs (a_i, b_i) for which $x_i \notin X_M$.

Because $|Y| = \mathcal{O}(|T|^2)$, we also have that $|X_M| = |M| = \mathcal{O}(|T|^2)$, which in turn means that $|P'| = \mathcal{O}(|T|^2)$. Since $T' = T$, this gives us that G' has $\mathcal{O}(|T|^2)$ vertices and therefore $\mathcal{O}(|T|^3)$ edges.

This concludes the kernelization. First note that all steps of the kernelization (constructing B from G, finding a maximum matching in B and removing people pairs from G to obtain G') can be done in polynomial time. Now to show that this procedure is indeed a valid kernelization for PM, it remains to show that it provides a correct reduction.

To this end, we show that (G, c, f, d) is a YES-instance if and only if (G', c, f, d) is a YES-instance. Clearly, if (G', c, f, d) is a YES-instance, then so is (G, c, f, d), since G' is obtained from G by only removing people vertices. Then every matching in G' also exists in G, since G' is a subgraph of G. Moreover, since no task vertices were removed from G to obtain G', a matching that covers all task vertices in G' also does so in G.

Suppose now that (G, c, f, d) is a YES-instance, meaning that there is some matching M_G in G of cost $d^* \leq d$. If M_G only contains edges that are also present in G', then this matching also exists in G', making (G', c, f, d) trivially a YES-instance. So suppose M_G does not exist in G'. Then we will construct a matching $M_{G'}$ in G' with the same cost d^*.

To this end, we take a look at the graph B. Although B was just an auxiliary graph to construct G' (and is therefore not part of our resulting PM instance), it will be useful in constructing the matching $M_{G'}$. As explained above, the matching M_G in G could be encoded in B as some matching M_B. Let $Y' \subseteq Y$ denote the set of vertices in Y that are covered by this M_B. Each vertex in Y' represents an ordered combination of either one or two vertices that are matched to the same pair of people vertices by M_G.

Consider now again the maximum matching M in X from before and the set X_M of vertices in X covered by M. By Lemma 1, there exists a matching M'' in $B[X_M \cup Y]$ covering all the vertices in Y'. We use this matching M'' to construct $M_{G'}$. We do the following for every edge $e \in M''$:

- If e has endpoints $x_i \in X$ and $y_j \in Y_1$, then we add either $\{a_i, t_j\}$ or $\{b_i, t_j\}$ to $M_{G'}$, depending on which of these edges is present in G'. We know that at least one of these edges is present in G', because e exists in B and $x_i \in X_M$.
- If e has endpoints $x_i \in X$ and $y_{j_1,j_2} \in Y_2$, then we add the edges $\{a_i, t_{j_1}\}$ and $\{b_i, t_{j_2}\}$ to $M_{G'}$. Because e exists in B, both these edges are present in G and because $x_i \in X_M$, this edge also exists in G'.

Because M'' covers the same vertices in Y as M_B does, this construction of $M_{G'}$ leads to a matching in G', which matches the same combinations of vertices to the same pair of people vertices as M_G did in G. Not only does this make $M_{G'}$ a valid matching in G', but it also has the same cost d^* as M_G, making (G', c, f, d) a YES-instance. □

Of course, performing just the kernelization on a PM instance does not yet yield an answer to the problem. Combining it with a brute-force algorithm however, yields an exact algorithm for the problem as follows.

Corollary 2 *Any PM instance* $I = (G, c, f, d)$ *with* $G = (P + T, E)$ *can be solved in time* $2^{\mathcal{O}(|T|^3)} \cdot |I|^{\mathcal{O}(1)}$.

Proof Let $I = (G, c, f, d)$ be a PM-instance with $G = (P + T, E)$. To solve it in the desired running time, the first step would be to perform the kernelization from Proposition 1. This takes $|I|^{\mathcal{O}(1)}$ time and yields an equivalent instance (G', c, f, d) in which G has $\mathcal{O}(|T|^2)$ vertices and $\mathcal{O}(|T|^3)$ edges.

The second step is to solve the new instance using a brute-force algorithm. A very naive way in which this can be done is to iterate over all $2^{\mathcal{O}(|T|^3)}$ subsets of edges in G' and checking whether they are a matching such that all task vertices are covered at a cost of at most d. Doing this can of course be done in $|I|^{\mathcal{O}(1)}$ time, so the total time spent by this procedure is $2^{\mathcal{O}(|T|^3)} \cdot |I|^{\mathcal{O}(1)}$. □

4 Conclusion and Discussion

In this paper we have seen the introduction of the PAIRED MATCHING problem and explored its algorithmic complexity. Further exploration thereof has been done in the master thesis this work is based on [12], but plenty of open questions related to the problem remain.

While we have seen the problem to be NP-complete, not much is known about special cases of the problem being solvable in polynomial time. Some NP-complete graph problems become polynomial time solvable on planar graphs for example and one could investigate whether the same is true for PM. Another restriction of PM that could be considered is obtained by requiring the cost function f to be constant.

It could also be interesting to see whether improvements can be made to the algorithm presented in Corollary 2. Either of its two parts can be tackled. First, one could see whether the kernelization can be improved to yield an even smaller instance. If not, it may be possible to prove that any kernelization of PM must yield instances whose size are at least cubic in the number of task vertices, just like the kernelization we have provided in Proposition 1. Secondly, it could be investigated whether the follow-up algorithm could be improved over the naive brute-force algorithm presented here.

References

1. Bodlaender, H., Jansen, B., Kratsch, S.: Kernel bounds for path and cycle problems. Theor. Comput. Sci. **511**, 117–136 (2013). https://doi.org/10.1016/j.tcs.2012.09.006
2. Cook, S.: The complexity of theorem-proving procedures. In: Harrison, M., Banerji, R., Ullman, J (eds.) Proceedings of the 3rd Annual ACM Symposium on Theory of Computing, pp. 151–158. ACM (1971). https://doi.org/10.1145/800157.805047
3. Cygan, M., Fomin, F., Kowalik, L., Lokshtanov, D., Marx, D., Pilipczuk, M., Pilipczuk, M., Saurabh, S.: Parameterized Algorithms. Springer (2015). https://doi.org/10.1007/978-3-319-21275-3
4. Edmonds, J.: Paths, trees, and flowers. Can. J. Math. **17**, 449–467 (1965). https://doi.org/10.4153/cjm-1965-045-4
5. Fellows, M., Knauer, C., Nishimura, N., Ragde, P., Rosamond, F., Stege, U., Thilikos, D., Whitesides, S.: Faster fixed-parameter tractable algorithms for matching and packing problems. Algorithmica **52**(2), 167–176 (2008). https://doi.org/10.1007/s00453-007-9146-y
6. Gupta, S., Roy, S., Saurabh, S., Zehavi, M.: Quadratic vertex kernel for rainbow matching. Algorithmica **82**(4), 881–897 (2020). https://doi.org/10.1007/s00453-019-00618-0
7. Kano, M., Li, X.: Monochromatic and heterochromatic subgraphs in edge-colored graphs-a survey. Graphs Comb. **24**(4), 237–263 (2008). https://doi.org/10.1007/s00373-008-0789-5
8. Maalouly, N.E., Steiner, R.: Exact matching in graphs of bounded independence number. In: Szeider, S., Ganian, R., Silva, A (eds.) 47th International Symposium on Mathematical Foundations of Computer Science, MFCS 2022, August 22–26, 2022, Vienna, Austria, *LIPIcs*, vol. 241, pp. 46:1–46:14. Schloss Dagstuhl-Leibniz-Zentrum für Informatik (2022). https://doi.org/10.4230/LIPIcs.MFCS.2022.46
9. Moser, H., Thilikos, D.: Parameterized complexity of finding regular induced subgraphs. J. Discret. Algorithms **7**(2), 181–190 (2009). https://doi.org/10.1016/j.jda.2008.09.005

10. Papadimitriou, C., Yannakakis, M.: The complexity of restricted spanning tree problems. J. ACM **29**(2), 285–309 (1982). https://doi.org/10.1145/322307.322309
11. Stockmeyer, L., Vazirani, V.: NP-completeness of some generalizations of the maximum matching problem. Inf. Process. Lett. **15**(1), 14–19 (1982). https://doi.org/10.1016/0020-0190(82)90077-1
12. Verhaegh, R.: The parameterized complexity of a new matching problem: the paired matching problem. Master's thesis, Eindhoven University of Technology (2022). https://pure.tue.nl/ws/portalfiles/portal/292963340/Verhaegh_R.pdf
13. Yannakakis, M.: Node- and edge-deletion NP-complete problems. In: Lipton, R., Burkhard, W., Savitch, W., Friedman, E., Aho, A (eds.) Proceedings of the 10th Annual ACM Symposium on Theory of Computing, pp. 253–264. ACM (1978). https://doi.org/10.1145/800133.804355

Edge Contraction and Forbidden Induced Subgraphs

Hany Ibrahim and Peter Tittmann

Abstract Given a family of graphs $\mathcal{H}$, a graph G is $\mathcal{H}$-free if any subset of $V(G)$ does not induce a subgraph of G that is isomorphic to any graph in $\mathcal{H}$. We present sufficient and necessary conditions for a graph G such that G/e is $\mathcal{H}$-free for any edge e in $E(G)$. Thereafter, we use these conditions to characterize $2K_2$-free, C_4-free, C_5-free, and split graphs.

1 Introduction

A graph G is an ordered pair $(V(G), E(G))$ where $V(G)$ is a set of vertices and $E(G)$ is a set of 2-elements subsets of $V(G)$ called edges. Thus, any graph in this paper is simple. The set of all graphs is $\mathcal{G}$. The degree of a vertex v, denoted by $deg(v)$, is the number of edges incident to v. We denote the maximum degree of a vertex in a graph G by $\Delta(G)$. We call two vertices adjacent if there is an edge between them, otherwise, we call them nonadjacent. Moreover, the set of all vertices adjacent to a vertex v is called the *neighborhood* of v, which we denote by $N(v)$. On the other hand, the *closed neighborhood* of v, denoted by $N[v]$, is $N(v) \cup \{v\}$. Generalizing this to a set of vertices S, the neighborhood of S, denoted by $N(S)$, is defined by $N(S) := \bigcup_{v \in S} N(v) \setminus S$. Similarly the closed neighborhood of S, denoted by $N[S]$, is $N(S) \cup S$. Moreover, for a subset of vertices S, we denote the set of vertices in S that are adjacent to v by $N_S(v)$. Furthermore, we write v is adjacent to S to mean that $S \subseteq N(v)$ and v is adjacent to exactly S to mean that $S = N(v)$.

A set of vertices S is *independent* if there is no edge between any two vertices in S. We call a set S *dominating* if $N[S] = V(G)$. A subgraph H of a graph G is a graph where $V(H) \subseteq V(G)$ and $E(H) \subseteq E(G)$. An *induced* graph $G[S]$ for a given

H. Ibrahim (✉)
KIT university, Karlsruhe, Germany
e-mail: hany.ibrahim@kit.edu

P. Tittmann
University of Applied Sciences Mittweida, Mittweida, Germany
e-mail: peter@hs-mittweida.de

A. Brieden et al. (eds.), *Graphs and Combinatorial Optimization: from Theory to Applications*, AIRO Springer Series 13,
https://doi.org/10.1007/978-3-031-46826-1_2

set $S \subseteq V$, is a subgraph of G with vertex set S and two vertices in $G[S]$ are adjacent if and only if they are adjacent in G. Two graphs G, H are *isomorphic* if there is a bijective mapping $f : V(G) \to V(H)$ where $u, v \in V(G)$ are adjacent if and only if $f(u), f(v)$ are adjacent in H. In this case we call the mapping f an isomorphism. Two graphs that are not isomorphic are called *non-isomorphic*. In particular, an isomorphism from a graph to itself is called *automorphism*. Furthermore, two vertices u, v are similar in a graph G if there is an automorphism that maps u to v. The set of all automorphisms of a graph G forms a group called the automorphism group of G, denoted by *Aut(G)*. The complement of a graph G, denoted by $\bar{G}$, is a graph with the same vertex set as $V(G)$ and two vertices in $\bar{G}$ are adjacent if and only if they are nonadjacent in G.

The *independence number* of a graph G, denoted by $\alpha(G)$, is the largest cardinality of an independent set in G. In this thesis, we write singletons $\{x\}$ just as x whenever the meaning is clear from the context. A vertex u is a *corner dominated* by v if $N[u] \subseteq N[v]$. Let $\mathcal{H}$ be a set of graphs. A graph G is called *$\mathcal{H}$-free* if there is no induced subgraph of G that is isomorphic to any graph in $\mathcal{H}$, otherwise, we say G is *$\mathcal{H}$-exist*.

By *contracting* the edge between u and v, we mean the graph constructed from G by adding a vertex w with edges from w to the union of the neighborhoods of u and v, followed by removing u and v. We denote the graph obtained from contracting uv by G/uv. If e is the edge between u and v, then we also denote the graph G/uv by G/e. Further, we call G/e a G-contraction. Finally, for notions not defined, please consult [2]. Additionally, we divide longer proofs into smaller claims, and we prove them only if their proofs are not apparent.

For a graph invariant c, a graph G, and a G-contraction H, the question of how $c(G)$ differs from $c(H)$ is investigated for different graph invariants. For instance, how contracting an edge in a graph affects its k-connectivity. Hence, the intensively investigated ([13]) notion of *k-contractible* edges in a k-connected graph G is defined as the edge whose contraction yields a k-connected graph. Another instance is in the game Cops and Robber where a policeman and a robber are placed on two vertices of a graph in which they take turns to move to a neighboring vertex. For any graph G, if the policeman can always end in the same vertex as the robber, we call G *cop-win*. However, G is $CECC$ if it is not cop-win, but any G-contraction is cop-win. The characteristics of a $CECC$ graph are studied in [5].

A further instance is the investigation of the so-called *contraction critical edges*, with respect to independence number. That is an edge e in a graph G where $\alpha(G/e) \leq \alpha(G)$, studied in [18]. Furthermore, the case where c is the chromatic and clique number, respectively, has been investigated in [7, 16, 17].

In this article, we investigate the graph invariant H-free for a given set of graphs $\mathcal{H}$. In particular, we present sufficient and necessary conditions for a graph G such that any G-contraction is $\mathcal{H}$-free.

Let $\mathcal{H}$ be a set of graphs. The set of *elementary (minimal)* graphs in $\mathcal{H}$, denoted by elm*($\mathcal{H}$)*, is defined as $\{H \in \mathcal{H} :$ if $G \in \mathcal{H}$ and H is G-exist, then G is isomorphic to $H\}$. From the previous definition, we can directly obtain the following.

Proposition 1 *Let $\mathcal{H}$ be a set of graphs. Graph G is $\mathcal{H}$-free if and only if G is* $\text{elm}(\mathcal{H})$*-free.*

We call an $\mathcal{H}$-free graph G, *strongly $\mathcal{H}$-free* if any G-contraction is $\mathcal{H}$-free. Furthermore, an $\mathcal{H}$-exist graph G is a *critically $\mathcal{H}$-exist* if any G-contraction is $\mathcal{H}$-free. If we add any number of isolated vertices to a strongly H-free or critically H-exist graph, then we obtain a graph with the same property. Thus, from this section and forward, we exclude graphs having isolated vertices unless otherwise stated.

We conclude directly the following.

Proposition 2 *Let $\mathcal{H}$ be a set of graphs and G be a graph where G is neither critically $\mathcal{H}$-exist nor $\mathcal{H}$-free but not strongly $\mathcal{H}$-free. The graph G is $\mathcal{H}$-free if and only if any G-contraction is $\mathcal{H}$-free.*

Given a graph G and a set of graphs $\mathcal{H}$, we call G *$\mathcal{H}$-split* if there is a G-contraction isomorphic to a graph in $\mathcal{H}$. Furthermore, G is *$\mathcal{H}$-free-split* if G is $\mathcal{H}$-split and $\mathcal{H}$-free. Moreover, the set of all $\mathcal{H}$-free-split graphs, for a given $\mathcal{H}$, is denoted by $\text{fs}(\mathcal{H})$.

Proposition 3 *Let $\mathcal{H}$ be a set of graphs and G be a $\mathcal{H}$-free graph. Then G is strongly $\mathcal{H}$-free if and only if G is* $\text{fs}(\mathcal{H})$*-free.*

Proof Assume for the sake of contradiction that there exists a strongly $\mathcal{H}$-free graph G with an induced $\mathcal{H}$-free-split subgraph J. Consequently, there is an edge e in J such that J/e induces a graph in $\mathcal{H}$. As a result, G/e is $\mathcal{H}$-exist, which contradicts the fact that G is strongly $\mathcal{H}$-free.

In contrast, if G is an $\mathcal{H}$-free but not a strongly $\mathcal{H}$-free, then there is a set $U \subseteq V(G)$ such that there is an edge $e \in E(G[U])$ where G/e is $\mathcal{H}$-exist. Let U be a minimum set with such a property. Thus $G[U]$ is $\mathcal{H}$-free-split. □

From Propositions 2 and 3, we deduce the following.

Theorem 1 *Let $\mathcal{H}$ be a set of graphs and G be a* $\text{fs}(\mathcal{H})$*-free graph where G is not critically $\mathcal{H}$-exist. The graph G is $\mathcal{H}$-free if and only if any G-contraction is $\mathcal{H}$-free.*

Theorem 1 provides a sufficient and necessary condition that answers the question we investigate in this article, however, it translates the problem to determining characterizations for critically $\mathcal{H}$-exist and $\mathcal{H}$-free-split graphs for a set of graphs $\mathcal{H}$. In Sects. 1.1 and 1.2, we present some properties for these families of graphs.

1.1 The H-Split Graphs

Let H be a graph with $v \in V(H)$ and $N_H(v) = U \cup W$. The *splitting(H,v,U,W)* is the graph obtained from H by removing v and adding two vertices u and w where $N_H(u) = U \cup \{w\}$ and $N_H(w) = W \cup \{u\}$. Furthermore, $splitting(H, v)$ is the set of all graphs for any possible U and W. Moreover, $splitting(H)$ is the union of the $splitting(H, v)$ for any vertex $v \in V(H)$. Given a set of graphs $\mathcal{H}$, $splitting(\mathcal{H})$ is the union of the splittings of every graph in $\mathcal{H}$.

Theorem 2 *For a graph G and a set of graphs $\mathcal{H}$, G is an $\mathcal{H}$-split if and only if $G \in splitting(\mathcal{H})$.*

Proof Let G be an $\mathcal{H}$-split. Hence there is a graph $H \in \mathcal{H}$ such that G is H-split. Thus, there are two vertices $u, w \in V(G)$ such that G/uw is isomorphic to H. Let $x := V(G/uw) - V(G)$, then $N_{G/uw}(x) = (N_G(u) \cup N_G(w)) \setminus \{u, w\}$. As a result, $G \in splitting(H, x)$. Consequently, $G \in splitting(\mathcal{H})$.

Conversely, let $G \in splitting(\mathcal{H})$. Hence there is a graph $H \in \mathcal{H}$ such that $G \in splitting(H)$. Thus, there are two adjacent vertices $u, w \in V(G)$ such that $G/uw \cong H$. Thus, G is $\mathcal{H}$-split. □

For a set of graphs $\mathcal{H}$ and using Theorem 2, we can use $splitting(\mathcal{H})$ to construct all $\mathcal{H}$-split graphs, consequently $\mathcal{H}$-free-split graphs.

Proposition 4 *In a graph G, let $u, v \in V(G)$. If u is similar to v, then $splitting(G, u) = splitting(G, v)$.*

By the previous proposition, for a graph H, the steps to construct the H-free-split graphs are:

- Let π be the partition of $V(H)$ induced by the orbits generated from $Aut(H)$;
- for every orbit $o \in \pi$, we choose a vertex $v \in o$; and
- construct $splitting(H, v)$.

The proofs of Propositions 5 to 8 are direct. Thus, we skip them and leave them for the interested reader.

Proposition 5 *Let G be a graph, v a vertex in $V(G)$ where $N_G(v) = U \cup W$. If $U = N_G(v)$ or $W = N_G(v)$, then $splitting(G, v, U, W)$ is not G-free-split.*

Proposition 6 *Let G be a graph and v a vertex in $V(G)$. If $deg(v) = 1$, then $splitting(G, v)$ contains no G-free-split graph.*

Proposition 7 *If G is a path, then $splitting(G)$ contains no G-free-split graph.*

Proposition 8 *If G is a C_n for an integer $n \geq 3$, then the G-free-split is C_{n+1}.*

1.2 Critically H-Exist Graphs

Theorem 3 *Let G be a graph and $\mathcal{H}$ be a set of graphs. If G is a critically $\mathcal{H}$-exist, then for any $S \subseteq V(G)$ such that $G[S]$ is isomorphic to a graph in $\mathcal{H}$, the followings properties hold:*

1. *$V(G) \setminus S$ is independent and*
2. *there is no corner in $V(G) \setminus S$ that is dominated by a vertex in S.*

Proof 1. For the sake of contradiction, assume there is a $S \subseteq V(G)$ such that $G[S]$ is isomorphic to a graph $H \in \mathcal{H}$ but $V(G) \setminus S$ is not independent. Hence, there are two vertices $u, v \in V(G) \setminus S$ where u and v are adjacent. Consequently, $G/uv[S]$ is isomorphic to H, which contradicts the fact that G is a critically $\mathcal{H}$-exist.
2. Since $V(G) \setminus S$ is independent, the neighborhood of any vertex in $V(G) \setminus S$ is a subset of S. For the sake of contradiction, assume that there is a corner $u \in V(G) \setminus S$ that is dominated by $v \in S$. However, $G/uv[S]$ is isomorphic to a graph $H \in \mathcal{H}$, which contradicts the fact that G is a critically $\mathcal{H}$-exist.

□

Corollary 1 *Let G be a critically $\mathcal{H}$-exist graph for a set of graphs $\mathcal{H}$. If S is a vertex set that induces a graph in $\mathcal{H}$, then no vertex in $V(G) \setminus S$ is adjacent to exactly one vertex, two adjacent vertices, three vertices that induce either P_3 or C_3, or a vertex with degree $|V(G)| - 1$.*

Let G be a graph with adjacent vertices u, v, and $\{w\} := V(G/uv) \setminus V(G)$. We define the mapping $f : 2^{V(G)} \rightarrow 2^{V(G/uv)}$ as follows:

$$f(S) = \begin{cases} S & \text{if } u, v \notin S, \\ (S \cup \{w\}) \setminus \{u, v\} & \text{otherwise.} \end{cases}$$

Let S be a vertex set such that $G[S]$ is isomorphic to a given graph H. We call an edge uv, H-critical for S if $G/uv[f(S)]$ is non-isomorphic to H. Furthermore, we call the edge uv H-critical in G if for any vertex subset S that induces H, uv is H-critical for S.

Theorem 4 *Let G be a graph and $S \subseteq V(G)$ where H is the graph induced by S in G. For any edge $uv \in E(G)$, uv is H-critical for S if and only if*

1. *$u, v \in S$ or*
2. *$u \in V(G) \setminus S$, $v \in S$, and u is not a corner dominated by v in the subgraph $G[S \cup \{u\}]$.*

Proof 1. If $u, v \in S$, then $|f(S)| < |S|$. Thus, $G/uv[f(S)]$ is non-isomorphic to H.
2. Let $u \in V(G) \setminus S$, $v \in S$, and u is not a corner dominated by v in the subgraph $G[S \cup \{u\}]$. Additionally, let $w \in N_S(u)$ but $w \notin N_S(v)$. In G/uv, let $x := V(G/uv) \setminus V(G)$. Clearly, x is adjacent to any vertex in $N_S(v) \cup \{w\}$. Hence, the size of $G/uv[f(S)]$ is larger than that of $G[S]$. Thus, $G/uv[f(S)]$ is non-isomorphic to H.

Conversely, if none of the conditions in the theorem hold, then one of the following holds:

1. both u and v are not in S, or
2. $u \in V(G) \setminus S, v \in S$, and u is a corner dominated by v in the subgraph $G[S \cup \{u\}]$.

In both cases, $G[S] \cong G/uv[f(S)] \cong H$. Consequently, uv is not H-critical for S. □

In the following Section, we present examples on using Theorem 1 by using it to characterize the special graph classes $2K_2$-free, C_4-free, C_5-free, and split graphs.

2 Special Graph Classes

2.1 The $2K_2$-Free Graphs

Different graphs families are $2K_2$-free graphs; for instance split, pseudo-split, threshold, and co-chordal graphs. Various graph invariants were studied for $2K_2$-free graphs, please consult [3, 4, 6, 8, 10]. The class of $2K_2$-free graphs has been characterized in different ways, see [14, 19].

We call an edge uv in a graph G *almost-dominating* if $V(G) \setminus N[\{u, v\}]$ induces edgeless graph.

Proposition 9 *A graph G is $2K_2$-free if and only if any edge in $E(G)$ is almost-dominating.*

Lemma 1 *Let G be a graph with a unique subset $S \subseteq V(G)$ such that $G[S]$ induces $2K_2$. If every edge e in $E(G)$ is e is $2K_2$-critical for S, then G is a critically $2K_2$-exist.*

Proof Let H be a G-contraction. Every edge e in $E(G)$ is $2K_2$-critical for S, then $V(G) \setminus S$ is independent set. Furthermore, every vertex in $V(G) \setminus S$ is adjacent to at least two nonadjacent vertices in S. In H, let $u \in V(G) \setminus f(S)$ and $v \in f(S)$. If u, v are adjacent, then uv is almost-dominating. Let $u \in f(S)$, then uv is almost-dominating. Hence, every edge in H is almost-dominating. Thus, G is a critically $2K_2$-exist. □

It is not hard to identify the $2K_2$-split graphs.

Proposition 10 *The graphs $P_2 \cup C_3$ and $P_2 \cup P_3$ are the only $2K_2$-split graphs.*

Clearly, both $P_2 \cup C_3$ and $P_2 \cup P_3$ are $2K_2$-exist. Thus, the following corollary follows directly.

Corollary 2 *There is no $2K_2$-free-split graph.*

Proposition 11 *The graphs in Fig. 1 are the only critically $2K_2$-exist graphs.*

Proof Through this proof, we assume that G is a critically $2K_2$-exist graph with $S = \{r, s, t, u\}$ such that $G[S]$ is isomorphic to $2K_2$, where rs and tu are edges in G. By Theorem 3, we note that $V(G) \setminus S$ is independent. Thus, any vertex in $V(G) \setminus S$ is adjacent to vertices only in S. By Corollary 1, if $v \in V(G) \setminus S$, then neither $|N(v)| = 1$ nor v is adjacent to exactly two adjacent vertices.

Claim 11.1 *If $v, w \in V(G) \setminus S$ such that $|N(v)| = |N(w)| = 2$ while $N(v) \cap N(w) = \phi$, then G is isomorphic to H_1.* □

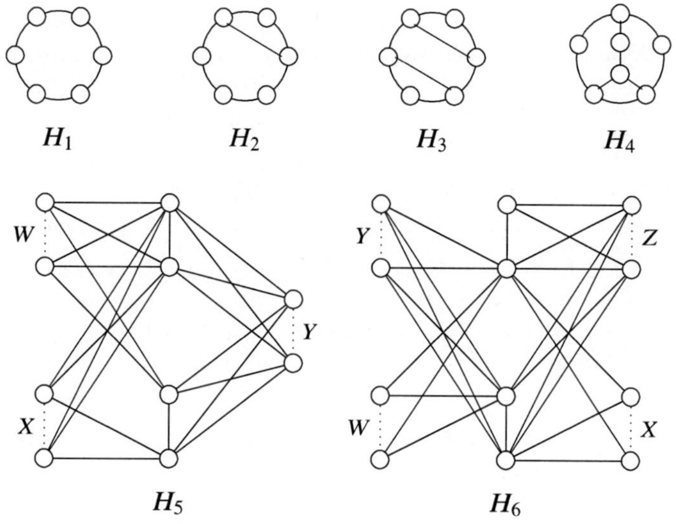

Fig. 1 Critically $2K_2$-exist graphs

Proof W.l.o.g., let $N(v) = \{r, u\}$ and $N(w) = \{s, t\}$. We will show that $V(G) = S \cup \{v, w\}$. For the sake of contradiction, assume that there is a vertex $x \in V(G) \setminus S$. Thus, x is adjacent to at least one vertex in S. W.l.o.g., let x be adjacent to r. In G/rx, $f(\{s, u, v, w\})$ induces $2K_2$, which contradicts the fact that G is a critically $2K_2$-exist. □

Claim 11.2 *If $v, w, x \in V(G) \setminus S$ such that $|N(v)| = |N(w)| = 2$, $|N(x)| = 3$ while $N(v) = N(w)$, then $N(v) \subset N(x)$.* □

Proof W.l.o.g., let $N(v) = N(w) = \{r, u\}$. For the sake of contradiction and w.l.o.g., assume $N(x) = \{r, s, t\}$. In G/rw, $f(\{s, u, v, x\})$ induces $2K_2$, which contradicts the fact that G is a critically $2K_2$-exist. □

Claim 11.3 *If $v, w, x \in V(G) \setminus S$ such that $|N(v)| = |N(w)| = 2$ while $|N(v) \cap N(w)| = 1$ and $|N(x)| = 3$ where $N(v) \cap N(w) \cap N(x) = \phi$, then G is isomorphic to H_4.* □

Proof W.l.o.g., let $N(v) = \{r, u\}$, $N(w) = \{r, t\}$, and $N(x) = \{s, t, u\}$. For the sake of contradiction, assume that there is a vertex $y \in V(G) \setminus S$. Hence, y is adjacent to at least one vertex in S. If y is adjacent to s (or u), then $f(\{r, t, v, x\})$ induces $2K_2$ in G/sy (or G/uy), which contradicts the fact that G is a critically $2K_2$-exist. Moreover, if y is adjacent to t, then $f(\{r, u, w, x\})$ induces $2K_2$ in G/ty, which contradicts the fact that G is a critically $2K_2$-exist. □

Claim 11.4 *If $v, w, x \in V(G) \setminus S$ such that $|N(v)| = |N(w)| = 2$ while $|N(v) \cap N(w)| = 1$ and $|N(x)| = 3$, then either $N(v) \cup N(w) = N(x)$ or $N(v) \cap N(w) \cap N(x) = \phi$ and G is isomorphic to H_4.* □

Proof By Claim 11.3, if $N(v) \cap N(w) \cap N(x) = \phi$, then G is isomorphic to H_4. If $N(v) \cup N(w) = N(x)$, then we are done. As a result, and w.l.o.g, let $N(v) = \{r, u\}$ and $N(w) = \{r, t\}$. Assume for the sake of contradiction that $N(x) = \{r, s, t\}$. However, $f(\{s, u, v, x\})$ induces $2K_2$ in G/rw, which contradicts the fact that G is a critically $2K_2$-exist. □

Claim 11.5 *If $v, w \in V(G) \setminus S$ such that $|N(v)| = |N(w)| = 3$, while $N(v) \cap N(w)$ consists of two nonadjacent vertices in S, then G is isomorphic to H_3.* □

Proof W.l.o.g., let $N(v) = \{r, t, u\}$ and $N(w) = \{r, s, t\}$. For the sake of contradiction, assume that there is a vertex $x \in V(G) \setminus S$. If x is adjacent to r (or t), then $f(\{s, u, v, w\})$ induces $2K_2$ in G/rx (or G/tx), which contradicts the fact that G is a critically $2K_2$-exist graph. Thus, $N(x) = \{s, u\}$, however, $f(\{s, t, v, x\})$ induces $2K_2$ in G/rw, which contradicts the fact that G is a critically $2K_2$-exist. □

Claim 11.6 *If $v, w, x \in V(G) \setminus S$ such that $|N(v)| = |N(w)| = 3$, while $N(v) \cap N(w)$ is two adjacent vertices in S, then $|N(x)| \neq 2$.* □

Proof W.l.o.g., Let $N(v) = \{r, t, u\}$ and $N(w) = \{s, t, u\}$. W.l.o.g and for the sake of contradiction, assume that $N(x) = \{r, u\}$. However, $f(\{r, t, w, x\})$ induces $2K_2$ in G/uv, which contradicts the fact that G is a critically $2K_2$-exist. □

By Claims 11.1 to 11.6, the possible critically $2K_2$-exist graphs are those presented in Fig. 1 whose proofs of being critically $2K_2$-exist for H_1, H_2, H_3, and H_4 are straightforward.

Claim 11.7 *The graph H_5 in Fig. 1 is a critically $2K_2$-exist.* □

Proof Graph H_5 in Fig. 1 is isomorphic to a graph G that contains a vertex subset $S = \{r, s, t, u\}$, where $G[S]$ is isomorphic to a $2K_2$ and $rs, tu \in E(G)$. Moreover, $V(G) = S \cup W \cup X \cup Y$, such that $N(w \in W) = \{r, s, t\}$, $N(x \in X) = \{r, s, u\}$, $N(y \in Y) = \{r, s, t, u\}$, and $|W|, |X|, |Y| \geq 0$.

We note that S is the only vertex set inducing $2K_2$ in G. Moreover, every edge in $E(G)$ is $G[S]$-critical for S. Thus, and by Lemma 1, H_5 in Fig. 1 is a critically $2K_2$-exist. □

Claim 11.8 *Graph H_6 in Fig. 1 is a critically $2K_2$-exist.* □

Proof Graph H_6 in Fig. 1 is isomorphic to a graph G that contains a vertex subset $S = \{r, s, t, u\}$ where $G[S]$ is isomorphic to a $2K_2$ and $rs, tu \in E(G)$. Moreover, $V(G) = S \cup W \cup X \cup Y \cup Z$, such that $N(w \in W) = \{s, t\}$, $N(x \in X) = \{s, u\}$, $N(y) = \{s, t, u\}$, $N(z) = \{r, s, t, u\}$, and $|W|, |X|, |Y|, |Z| \geq 0$.

We note that S is the only vertex set inducing $2K_2$ in G. Moreover, every edge in $E(G)$ is $G[S]$-critical for S. Thus, and by Lemma 1, H_6 in Fig. 1 is a critically $2K_2$-exist. □

By Claims 11.7 and 11.8, the proof is complete. □

By Theorem 1, Corollary 2, and Proposition 11, we obtain the following.

Theorem 5 *Let G be a graph that is non-isomorphic to any graph in Fig. 1. The graph G is $2K_2$-free if and only if any G-contraction is $2K_2$-free.*

2.2 The C_4-Free Graphs

The C_4-split graphs can be easily recognized as follows:

Proposition 12 *The graphs in Fig. 2 are the only C_4-split graphs.*

The remaining graphs presented in Fig. 2 are obviously C_4-exist, which immediately provides the following result.

Corollary 3 *C_5 is the only C_4-free-split graph.*

Proposition 13 *The graphs in Fig. 3 are the only critically C_4-exist graphs.*

Proof Through this proof, we assume that G is a critically C_4-exist graph with $S = \{r, s, t, u\}$ such that $G[S]$ is isomorphic to C_4 where r and t are adjacent to both s and u. By Theorem 3, we note that $V(G) \setminus S$ is independent. Thus, any vertex in $V(G) \setminus S$ is adjacent to vertices only in S. By Corollary 1, if $v \in V(G) \setminus S$, then v is nonadjacent to exactly: one vertex, two adjacent vertices, or three vertices.

Let $v \in V(G) \setminus S$ such that $|N(v)| = 2$. W.l.o.g, assume that $N(v) = \{r, t\}$. Let $w \in V(G) \setminus S$, however, if w is adjacent to s (or u), then in G/sw, $f(\{r, t, u, v\})$ induces C_4, which contradicts the fact that G is a critically C_4-exist. Thus, if $v \in V(G) \setminus S$ such that $|N(v)| = 2$, then G is isomorphic to a graph in H_1.

Let $v, w, x \in V(G) \setminus S$ such that $|N(v)| = |N(w)| = 4$. Hence, $|N(x)| = 4$, however, in G/rv, $f(\{s, u, w, x\})$ induces C_4, which contradicts the fact that G is a critically C_4-exist. Thus, if there is a vertex outside S that is adjacent to every vertex in S then G is isomorphic to either H_2 or H_3.

Consequently, the possible critically C_4-exist graphs are those presented in Fig. 3 whose proofs, of being critically C_4-exist, are straightforward, which completes the proof. □

By Theorem 1, Corollary 3, and Proposition 13, we obtain the following.

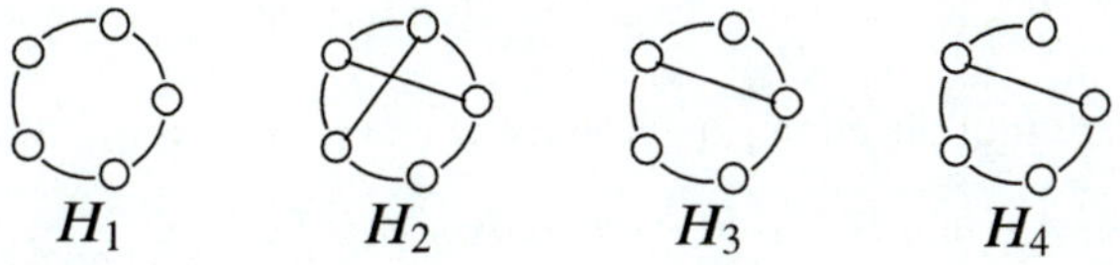

Fig. 2 C_4-split graphs

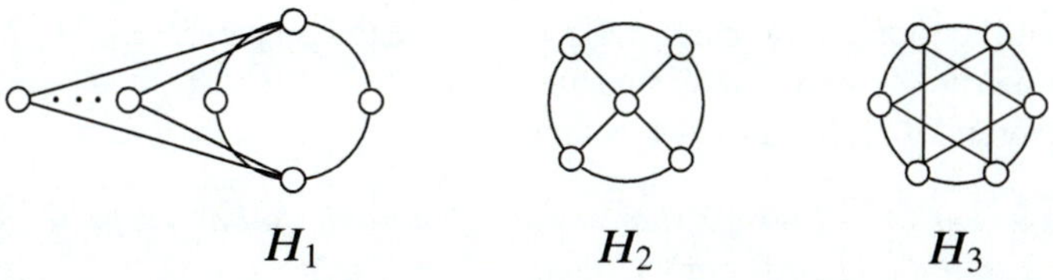

Fig. 3 Critically C_4-exist graphs

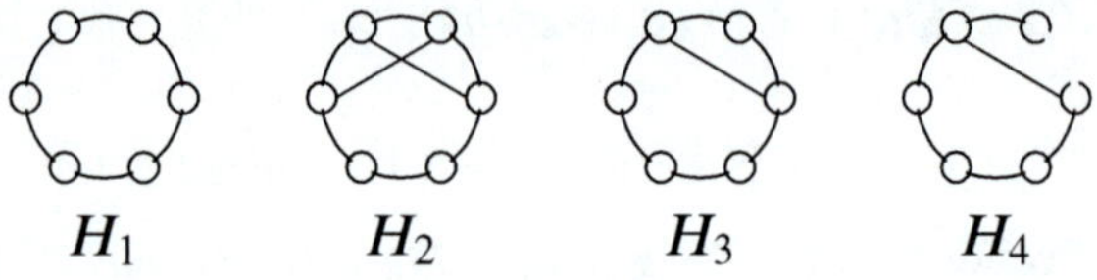

Fig. 4 C_5-split graphs

Theorem 6 *Let G be a C_5-free graph that is non-isomorphic to any graph in Fig. 3. The graph G is C_4-free if and only if any G-contraction is C_4-free.*

2.3 The C_5-Free Graphs

It is not hard to identify the C_5-split graphs as follows.

Proposition 14 *The graphs in Fig. 4 are the only C_5-split graphs.*

The remaining graphs presented in Fig. 4 are obviously C_5-exist, which immediately provides the following result.

Corollary 4 *C_6 is the only C_5-free-split graph.*

In similar way to the proof of Proposition 13, we can obtain the following:

Proposition 15 *The graphs in Fig. 5 are the only critically C_5-exist graphs.*

By Theorem 1, Corollary 4, and Proposition 15, we obtain the following.

Theorem 7 *Let G be a C_6-free graph that is non-isomorphic to any graph in Fig. 5. The graph G is C_5-free if and only if any G-contraction is C_5-free.*

2.4 Split Graphs

Split graphs were introduced in [9] and were characterized as follows:

Theorem 8 *[9] A graph G is split if and only if G is $\{2K_2, C_4, C_5\}$-free.*

Thus, we call a graph that is $\{2K_2, C_4, C_5\}$-exist *non-split* graph. Additionally, split graphs have been characterized in [9] as chordal graphs whose complements are also chordal. Furthermore, it was characterized by its degree sequences in [11]. Moreover, further properties of split graphs are studied in [1, 12, 15].

By Theorems 1, 5, 6, and 7, we obtain:

Theorem 9 *Let G be a graph that is non-isomorphic to any graph in Fig. 6. The graph G is split if and only if any G-contraction is split.*

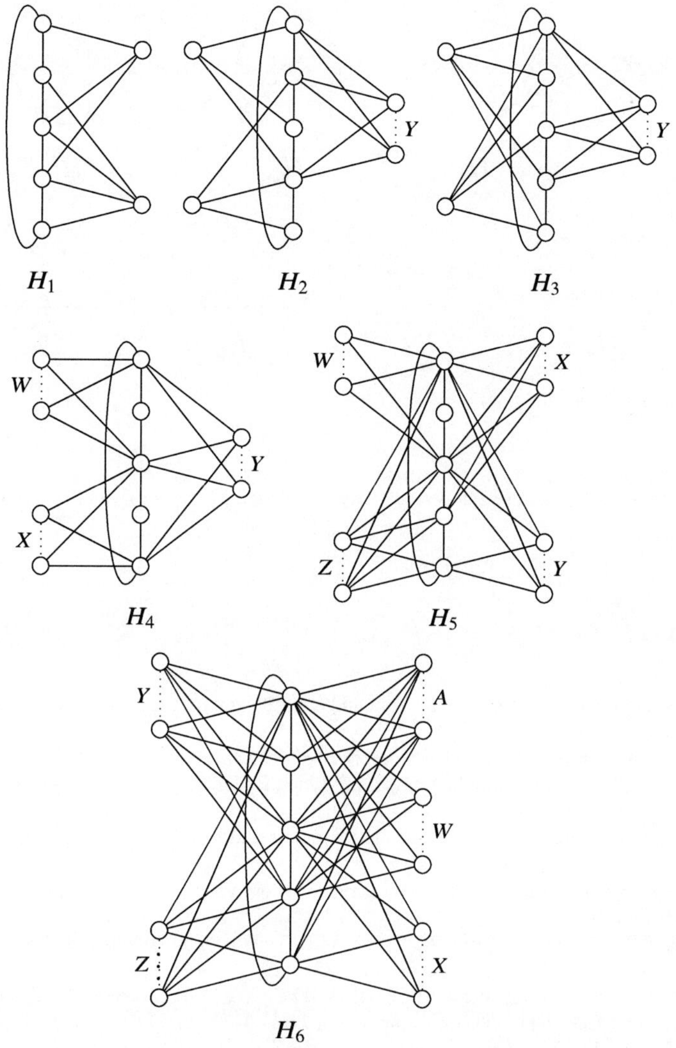

Fig. 5 Critically C_5-exist graphs

The class of split graphs is a closed class under edge contraction. The definition of a split graph implies that by contraction of an arbitrary edge in a split graph leads to another split graph. So the contribution of Theorem 9 is in listing the critically non-split graphs.

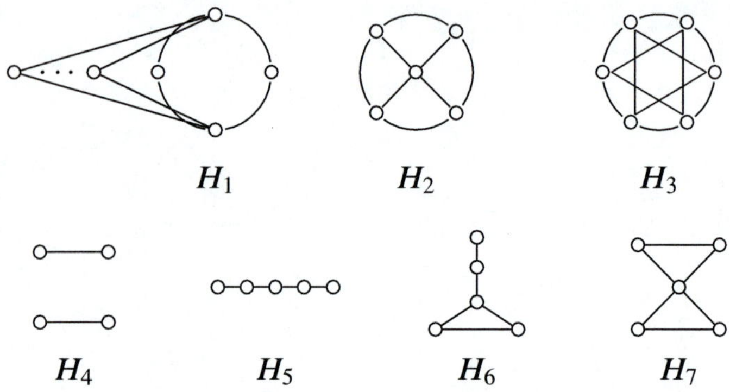

Fig. 6 Critically non-split graphs

Acknowledgements The research presented here is funded by the European Social Fund (ESF).

References

1. Bertossi, A.A.: Dominating sets for split and bipartite graphs. Inf. Process. Lett. **19**(1), 37–40 (1984)
2. Bondy, J.A., Murty, U.S.: Graph Theory. Springer (2000)
3. Brause, C., Randerath, B., Schiermeyer, I., Vumar, E.: On the chromatic number of $2K_2$-free graphs. Discret. Appl. Math. **253**, 14–24 (2019)
4. Broersma, H., Patel, V., Pyatkin, A.: On toughness and hamiltonicity of $2K_2$-free graphs. J. Graph Theory **75**(3), 244–255 (2014)
5. Cameron, B., Fitzpatrick, S.: Edge contraction and cop-win critical graphs. Australas. J. Comb. **63**, 70–87 (2015)
6. Chung, F.R., Gyárfás, A., Tuza, Z., Trotter, W.T.: The maximum number of edges in $2K_2$-free graphs of bounded degree. Discret. Math. **81**(2), 129–135 (1990)
7. Diner, Ö.Y., Paulusma, D., Picouleau, C., Ries, B.: Contraction and deletion blockers for perfect graphs and h-free graphs. Theor. Comput. Sci. **746**, 49–72 (2018)
8. El-Zahar, M., Erdős, P.: On the existence of two non-neighboring subgraphs in a graph. Comb. **5**(4), 295–300 (1985)
9. Foldes, S., Hammer, P.L.: Split graphs. In: Proceedings of the 8th Southeastern Conference on Combinatorics, Graph Theory, and Computing, Baton Rouge 1977, pp. 311–315 (1977)
10. Golan, G., Shan, S.: Nonempty intersection of longest paths in 2k_2-free graphs (2016). arXiv:1611.05967
11. Hammer, P.L., Simeone, B.: The splittance of a graph. Comb. **1**(3), 275–284 (1981)
12. Kratsch, D., Lehel, J., Müller, H.: Toughness, hamiltonicity and split graphs. Discret. Math. **150**(1), 231–246 (1996)
13. Kriesell, M.: A survey on contractible edges in graphs of a prescribed vertex connectivity. Graphs Comb. **18**(1), 1–30 (2002)
14. Meister, D.: Two characterisations of minimal triangulations of $2K_2$-free graphs. Discret. Math. **306**(24), 3327–3333 (2006)
15. Merris, R.: Split graphs. Eur. J. Comb. **24**(4), 413–430 (2003)

16. Paulusma, D., Picouleau, C., Ries, B.: Reducing the clique and chromatic number via edge contractions and vertex deletions. In: International Symposium on Combinatorial Optimization, pp. 38–49. Springer (2016)
17. Paulusma, D., Picouleau, C., Ries, B.: Critical vertices and edges in H-free graphs. Discret. Appl. Math. **257**, 361–367 (2019)
18. Plummer, M.D., Saito, A.: A note on graphs contraction-critical with respect to independence number. Discret. Math. **325**, 85–91 (2014)
19. Dhanalakshmi, S., N. S. and V. Manogna,: On $2K_2$-free graphs. Int. J. Pure Appl. Math. **109**(7), 167–173 (2016)

Exact Approaches for the Connected Vertex Cover Problem

Manuel Aprile

Abstract Given a graph G, the Connected Vertex Cover problem (CVC) asks to find a minimum cardinality vertex cover of G that induces a connected subgraph. This paper describes some approaches to solve the CVC problem exactly. First, we give compact mixed-integer extended formulations for CVC: these are the first formulations proposed for this problem, have a small number of extra variables and can be easily adapted to variations of the problem such as Tree Cover. Second, we describe a simple branch and bound algorithm for the CVC problem. Finally, we implement our algorithm and compare its performance against our best extended formulation: contrary to what usually happens for the classical Vertex Cover problem, our formulation outperforms the branch and bound algorithm.

1 Introduction

Given a graph $G = (V, E)$, a subset of vertices $C \subseteq V$ is a *vertex cover* of G if every edge of G has at least one endpoint in C. The problem of finding a vertex cover of minimum cardinality in a graph is equivalent to finding a maximum stable set (or a maximum clique in the complement graph) and is one of the best studied problems in theoretical computer science. In this paper we study one of the most popular variants of the minimum Vertex Cover (VC) problem, where we aim at finding a minimum *connected vertex cover* (CVC): i.e., we additionally require the subgraph $G[C]$ induced by C to be connected. We call this the CVC problem.

The CVC problem has applications in wireless network design, where one aims at placing relay stations on the network so that they cover all transmission links (the edges of the network) and are all connected to each other.

Similarly to the VC problem, the CVC problem is NP-hard [16] and admits a polynomial-time 2-approximation algorithm [25]. On the other hand, the CVC problem is NP-hard even if the input graph is restricted to be bipartite [14]: this is sur-

M. Aprile (✉)
Mathematics Department, Universitá degli studi di Padova, Via Trieste 63, 35121 Padova, Italy
e-mail: manuel.aprile@unipd.it

A. Brieden et al. (eds.), *Graphs and Combinatorial Optimization: from Theory to Applications*, AIRO Springer Series 13,
https://doi.org/10.1007/978-3-031-46826-1_3

prising as Vertex Cover is polynomially solvable for bipartite graphs, as, thanks to the famous König–Egeváry Theorem, it amounts to finding a maximum matching.

The CVC problem has received attention especially from the point of view of parameterized algorithms [17, 23] and approximation algorithms [10, 13, 25]. An aspect that did not receive much attention is that of solving the CVC problem in practice: moreover, prior to this paper there were no mathematical programming formulations for the problem. Such formulations are usually easy to implement and are flexible to the addition of extra constraints to the problem, an advantage for real-world applications. Different from the CVC problem, there is a wealth of methods for solving the VC problem, the most effective being branch and bound algorithms (see [27] for a survey), and there are many linear and non-linear formulations for VC and the related maximum clique and maximum stable set problems [5, 20, 24].

A key feature of the CVC problem that we exploit in this paper is that its constraints can be modelled as linear constraints from two polytopes: the vertex cover polytope and the spanning tree polytope. Both are well-studied polytopes for which a large number of *extended formulations* is known [5, 6, 8, 15, 21, 26]: those are formulations where extra variables are used, other than the variables of the original polytope, in order to limit the number of inequalities.

In this paper we aim at partially filling the gap between VC and CVC by proposing mixed-integer extended formulations for the CVC problem. Our main contribution is a mixed integer formulation for the CVC problem with a relatively small number of variables (linear in the number of edges of the input graph). The formulations we propose also lend themselves to modelling related problems as the Tree Cover problem [9] (see Sect. 5). As an additional contribution, we also describe a simple branch and bound algorithm for CVC, by modifying a standard algorithm for the maximum stable set problem. Finally, we perform numerical experiments to compare the various approaches. In our experiments, the proposed mixed-integer formulation solves the problem much faster than the branch and bound algorithm. This is interesting since, for the general Vertex Cover problem, combinatorial algorithms usually outperform linear formulations.

The paper is organized as follows: this introduction terminates with Sect. 1.1, which gives some basic terminology and notation; in Sect. 2 we give our formulations for CVC and prove their correctness; the branch and bound algorithm is described in Sect. 3; numerical experiments are given in Sect. 4; finally, we conclude with some further research directions in Sect. 5.

1.1 Preliminaries

Throughout the paper we let $G = (V, E)$ be a connected graph. This is natural because, ignoring exceptions such as isolated vertices, only connected graphs admit connected vertex covers. Given a set $U \subseteq V$, we let $G[U] = (U, E(U))$ be the subgraph induced by U, where $E(U) = \{(u, v) \in E : u, v \in U\}$. Set U is *stable* if the $G[U]$ does not contain any edge. Clearly, a subset $U \subseteq V$ is a vertex cover if and only if its complement $V \setminus U$ is stable. Hence, solving the CVC problem is equivalent to finding the maximum stable set S such that the graph $G \setminus S$ obtained by removing S is connected. Finally, a subgraph of G is a *spanning tree* of G if it is a tree and contains all vertices of G: we usually identify a spanning tree with a set of edges $F \subseteq E$.

For sets $U \subseteq A$, we denote by $\chi^U \in \{0, 1\}^A$ the *incidence* vector of U, which satisfies $\chi^U_v = 1$ if and only if $v \in U$. We will use incidence vectors for subsets of vertices, edges, or arcs in directed graphs. For a vector $x \in \mathbb{R}^A$, we often write $x(U)$ to denote $\sum_{u \in U} x_u$.

2 Mixed-Integer Programming Formulations

A compact integer formulation of the Vertex Cover problem is well known: it suffices to use a variable x_v for each node v of our graph G, and ask that $x_u + x_v \geq 1$ for each edge uv of G. On the other hand, it is not trivial to come up with a formulation for CVC, and we do not know any formulation that only uses node variables. The reason behind this difficulty is that imposing connectedness in an induced subgraph is a difficult constraint to model. Notice that a graph is connected if and only if it admits a spanning tree. Hence to model connectedness we resort to the spanning tree polytope of G, denoted by STP(G), defined as the convex hull of the incidence vectors of all the spanning trees in G. The basic idea that underlies all the formulations in this section is to add edge variables to the node variables, and to impose that such edge variables model a spanning tree in the subgraph induced by our vertex cover. We first propose the following formulation, based on the classical linear description of STP(G) given by Edmonds [12].

$$P_{\text{stp}} = \Big\{ x \in \{0, 1\}^V \mid \exists\, y \in [0, 1]^E :$$

$$x_u + x_v \geq 1 \qquad \forall (u, v) \in E \qquad (1)$$

$$y(E(U)) \leq |U| - 1 \qquad \forall \emptyset \neq U \subseteq V \qquad (2)$$

$$y(E) = x(V) - 1 \qquad (3)$$

$$y_{uv} \leq x_u,\ y_{uv} \leq x_v \qquad \forall (u, v) \in E \Big\}. \qquad (4)$$

Lemma 1 *Let $G = (V, E)$ be a connected graph. Then $C \subseteq V$ is a CVC if and only if $(\chi^C, y) \in P_{\text{stp}}$ for some $y \in \mathbb{R}^E$.*

Proof If C is a CVC, then fix any spanning tree F of $G[C]$. Then χ^C clearly satisfies Constraints (1); moreover, setting $y = \chi^F$ can be easily seen to satisfy Constraints (2), (3), (4).

On the other hand, assume that $(\chi^C, y) \in P_{\text{stp}}$. Then $C \subseteq V$ is clearly a vertex cover. Moreover, (4) implies $y_e = 0$ for each $e \in E \setminus E(C)$, hence the projection y' of y to variables $E(C)$ is in the spanning tree polytope of $G[C]$, due to constraints

(2), (3) (notice that $x(V) - 1 = |C| - 1$). The spanning tree polytope of $G[C]$ is then non-empty, therefore $G[C]$ is connected. □

The description above has an exponential number of constraints. There are well known extended formulations of size $O(n^3)$ for the spanning tree polytope of an n-vertex graph [21, 26], and smaller extended formulations for special classes of graphs [8, 15]. Therefore, we would like to turn any formulation for the spanning tree polytope into a formulation for CVC. This can be done by going through the forest polytope of G, $\mathrm{STP}^{\downarrow}(G)$, defined as the convex hull of incidence vectors of forests of G. The same proof of Lemma 1 shows that a correct formulation for CVC can be obtained by replacing Constraints 2 in P_{stp} with $y \in \mathrm{STP}^{\downarrow}(G)$. Finally, it is well-known that one can obtain a formulation of $\mathrm{STP}^{\downarrow}(G)$ from one of $\mathrm{STP}(G)$, since $\mathrm{STP}^{\downarrow}(G) = \{x \in [0, 1]^E : \exists y \in \mathbb{R}^E : x \leq y, y \in \mathrm{STP}(G)\}$. While this approach does reduce the size of our CVC formulation from exponential to polynomial, it still yields too many extra variables to be practical. In the next section, we address this issue.

2.1 A Smaller Mixed-Integer Formulation

We now give a smaller formulation for the CVC problem, which makes use of a mixed-integer formulation for $\mathrm{STP}(G)$ with a small number of additional variables. We start by giving the formulation for $\mathrm{STP}(G)$, which builds on natural ideas that can be found, for instance, in [22]. Rather than spanning trees in undirected graphs, we focus on arborescences in directed graphs. Given our graph G, we simply bidirect each edge obtaining the directed graph $D = (V, A)$. Now, fix a "root" vertex $r \in V$. Recall that an r-arborescence of D is a subset of arcs $F \subseteq A$ such that, for every $v \in V \setminus \{r\}$, F contains exactly one directed path from r to v. Clearly, a description of the r-arborescences of D gives a description of the spanning trees of G by just ignoring the orientations (i.e. setting $y_{uv} = z_{uv} + z_{vu}$ for each edge uv). Moreover, since arborescences are rooted in r, we do not need arcs that point to r, and we simply delete them. Recall that $\delta^-(v)$ denotes the set of arcs of A pointing to v. Consider the following formulation:

$$Q_r = \Big\{ z \in \{0, 1\}^A \mid \exists\, d \in \mathbb{R}^V :$$

$$z(\delta^-(v)) = 1 \qquad \forall v \in V \setminus \{r\} \qquad (5)$$

$$d_v \geq n \cdot (z_{uv} - 1) + d_u + 1 \qquad \forall (u, v) \in A \qquad (6)$$

$$d_r = 0 \qquad (7)$$

$$z(A) = |V| - 1 \Big\}. \qquad (8)$$

Lemma 2 *Let $D = (V, A)$ be a directed graph, and $r \in V$ such that $\delta^-(r) = \emptyset$. Then $F \subseteq A$ is an r-arborescence of D if and only if $(\chi^F, d) \in Q_r$ for some $d \in \mathbb{R}^V$.*

Proof First, given an r-arborescence F, set d_v to the length of the (unique) path from r to v in F, for each $v \in V$. It is easy to check that all constraints are satisfied by (χ^F, d).

On the other hand, let $(z, d) \in Q_r$, with $z = \chi^F$. We first show that F, after ignoring orientations, does not contain cycles: suppose by contradiction that $C \subseteq F$ is a cycle with vertices $v_1, \ldots, v_k$, where for each $i = 1, \ldots, k$, $v_i v_{i+1} \in C$ or $v_{i+1} v_i \in C$ (where the sum is modulo k). For any $uv \in C$, we have that $d_v \geq d_u + 1$ by (6): this implies that C cannot be a directed cycle. In particular, if v is the vertex of C with d_v minimum, then there are two arcs of C pointing to v: but this is in contradiction with Constraint (5), if $v \neq r$, and with $\delta^-(r) = \emptyset$ otherwise.

Now, we have $|F| = |V| - 1$ by (8). This, the absence of cycles, and Constraint (5), guarantees that F is an r-arborescence of D. □

One could turn Q_r into a formulation for the forest polytope of G and obtain a formulation for the CVC problem, as described in the previous section. However, it is not clear how to do this without adding additional variables: the issue is the choice of the root r, which does not need to be connected to the other vertices in a forest. Instead, we are able to limit the number of variables by exploiting the fact that, for any edge uv of G, at least one of u, v has to be picked in our vertex cover. Hence, we choose a "main" root vertex r, and another root r_1, adjacent to r, that we can use as a root when r is not in our vertex cover. We consider the following directed version $D = (V, A)$ of our graph $G = (V, E)$: fix $r, r_1 \in V$ with $rr_1 \in E$, turn every edge $vr \in E$ into a directed arc from r to v, turn every edge $vr_1 \in E$ with $v \neq r$ into a directed arc from r_1 to v, and bidirect each other edge. Notice that, in D, $\delta^-(r) = \emptyset$ and $\delta^-(r_1) = \{r\}$. Now, consider the following formulation:

$$P_{\text{arb}}(r, r_1) = \Big\{ x \in \{0, 1\}^V \;\; \exists\, z \in \{0, 1\}^A,\; d \in \mathbb{R}^V :$$

$$x_u + x_v \geq 1 \qquad \forall (u, v) \in A, \tag{9}$$

$$z(\delta^-(v)) = x_v \qquad \forall v \in V \setminus \{r, r_1\} \tag{10}$$

$$d_v \geq n \cdot (z_{uv} - 1) + d_u + x_v \qquad \forall (u, v) \in A \tag{11}$$

$$d_r = 0 \tag{12}$$

$$z(A) = x(V) - 1 \tag{13}$$

$$z_{uv} \leq x_u,\; z_{uv} \leq x_v \qquad \forall (u, v) \in A \Big\}. \tag{14}$$

Theorem 1 *Let $G = (V, E)$ be a connected graph, let $r, r_1 \in V$ with $(r, r_1) \in E$ and construct the directed graph $D = (V, A)$ as described above. Then $C \subseteq V$ is a CVC if and only if $(\chi^C, z, d) \in P_{\text{arb}}(r, r_1)$ for some z, d.*

Proof First, let $C \subseteq V$ be a CVC. We distinguish three cases.

1. $r \in C, r_1 \notin C$. Let F be any r-arborescence of $D[C]$, and set $x = \chi^C, z = \chi^F, d_v$ equal to the distance between r and v in F for $v \in C$, and $d_v = 0$ for $v \notin C$. Notice that $0 \leq d_v \leq n - 1$ holds for all $v \in V$. Now, (x, z, d) can be checked to satisfy all constraints of $P_{\text{arb}}(r, r_1)$: we only discuss Constraints (11). Let $(u, v) \in A$. If

$(u, v) \notin F$, the corresponding constraint is $d_v \geq -n + d_u + x_v$, which is trivially satisfied for any u, v as d_v is non-negative and the right-hand side is non-positive. Hence, suppose $(u, v) \in F$, hence $x_v = 1$. Then the constraint is $d_v \geq d_u + 1$, which is satisfied at equality by our choice of d.

2. $r_1 \in C, r \notin C$. We proceed similarly as in the previous case, choosing an r_1-arborescence F of $D[C]$ and setting $z = \chi^F$, d_v equal to the distance between r_1 and v in F for $v \in C$, and $d_v = 0$ for $v \notin C$. Then (x, z, d) can be checked to satisfy all constraints exactly as before.
3. $r, r_1 \in C$. Let F be an r-arborescence of $D[C]$ containing the arc rr_1 (notice that such an arborescence always exists). Set $z = \chi^F$, and set d as in the first case. Again, one checks that all constraints are satisfied.

Now, let $(\chi^C, z, d) \in P_{\text{arb}}(r, r_1)$, with $z = \chi^F$. In order to show that $G[C]$ is connected, we just need to show that F does not contain any cycle. We use the same argument as in the proof of Lemma 2, which we repeat for completeness. Assume that F contains a cycle C. C cannot be a directed cycle due to Constraints (11), hence C contains a vertex v with two incoming arcs. Constraint (10) implies that $v = r$ or $v = r_1$, but this contradicts the fact that $\delta^-(r) = \emptyset$, $\delta^-(r_1) = \{r\}$. □

3 A Branch & Bound Algorithm

In this section we describe a naive branch & bound algorithm to solve the CVC problem. For simplicity we follow the standard framework of branch & bound algorithms for the maximum stable set problem, see for instance [27]: instead of looking directly for a minimum vertex cover, we look for a stable set S^* of maximum size. The only difference with the classical setting is that we impose that S^* is feasible, where we call *feasible* a stable set S such that $G \setminus S$ is connected.

We now give an informal description of the algorithm, referring to Algorithm 1 for the pseudocode. To avoid recursion, a stack is used to store the nodes explored by the algorithm. Each node consists of a pair (S, U), where S is a feasible stable set and U is a set of candidate nodes that can be added to S. The idea is to explore the search space of all possible nodes while keeping a record of the best solution found so far, denoted by S^*: at each step, the current node (S, U) of the stack is either branched on, or pruned if we realize that it cannot produce a stable set larger than S^*. The pruning step is based on greedy coloring, as in the classical algorithm for the maximum stable set problem, exploiting the fact that any proper coloring of the complement of a graph gives an upper bound on its maximum stable set: in particular, the maximum stable set that the node can produce has size at most $|S| + \alpha(G[U]) \leq |S| + \chi(\bar{G}(U))$, and the latter term is estimated as the numbers of colors used in a greedy coloring (see Line 6). Branching is also performed as in the classical algorithm, but with a crucial difference: we select a vertex $v \in U$ and create nodes $(S, U \setminus \{v\})$ and $(S \cup \{v\}, U')$, where $U' \subseteq U \setminus \{v\}$ is obtained by removing from U all the neighbors of v and *all* *the cut-vertices*[1] of $G \setminus (S \cup \{v\})$ (see Line 11). This ensures that we only consider feasible stable sets.

Algorithm 1 Pseudocode of a basic branch & bound algorithm for CVC. Following the classical framework for maximum stable set algorithms, the algorithm finds the largest stable set S^* in G such that $G \setminus S^*$ is connected, and then outputs the corresponding vertex cover.

Input: A connected graph $G = (V, E)$

Output: A minimum-size CVC of G

1: $S^* \leftarrow \emptyset$
2: $C \leftarrow$ cut-vertices of G
3: $A \leftarrow [(\emptyset, V \setminus C)]$
4: **while** A non-empty **do**
5:   $(S, U) \leftarrow \text{pop}(A)$
6:   **while** U non-empty **and** $|S^*| < |S| +$ greedy_color($\bar{G}[U]$) **do**
7:     $v \leftarrow \text{pop}(U)$
8:     Append (S, U) to A
9:     $S \leftarrow S \cup \{v\}$
10:     $C \leftarrow$ cut-vertices of $G \setminus S$
11:     $U \leftarrow (U \cap \bar{N}(v)) \setminus C$
12:     **if** $|S| > |S^*|$ **then**
13:       $S^* \leftarrow S$
14:     **end if**
15:   **end while**
16: **end while**
17: **return** $V \setminus S^*$

We now argue that our algorithm is correct: most importantly, we need to show that removing cut-vertices as described above is enough to find the largest feasible stable set.

Theorem 2 *Let $G = (V, E)$ be a connected graph. Then Algorithm 1 on input G outputs a minimum CVC of G.*

Proof Equivalently, we will show that the set S^* output by the algorithm is the maximum feasible stable set of G. We say that a node (S, U) *contains* a feasible stable set S' if $S \subseteq S' \subseteq U$.

First, we claim that the starting node $(\emptyset, V \setminus C)$ contains all feasible stable sets, where C are the cut-vertices of G. Indeed, if u is a cut-vertex of G, and S a feasible stable set, S cannot contain u: if $u \in S$, we must have that $G \setminus \{u\}$ consists of two connected components G_1, G_2, and S contains the vertices of G_1 without loss of generality. But since G is connected, there is at least an edge between u and a vertex of G_1, a contradiction.

Now, it suffices to show that, whenever we branch on a node (S, U) obtaining two new nodes, any feasible stable set S' contained in (S, U) is contained in one of

---

[1] A vertex v of a connected graph G is a cut-vertex if its deletion disconnects G.

the new nodes. This implies that any feasible stable set is explored by the algorithm at some step, and concludes the proof.

The new nodes created are $(S, U \setminus \{v\})$ and $(S \cup \{v\}, U')$, where U' is defined in Line 11. Clearly, if $v \notin S'$, then S' is contained in node $(S, U \setminus \{v\})$ and we are done. On the other hand, if $v \in S'$, we only need to show that $S' \subseteq U'$. This follows since S' cannot contain any neighbor of v, or any cut-vertex of $G \setminus (S \cup \{v\})$, where the latter is proved by using the same argument as for the starting node. □

We conclude the section with some improvements to Algorithm 1 that can be implemented to increase performance (see next Section for the implementation details).

- Computing a strong upper bound reduces the number of branch and bound nodes, at the price of longer running time for each node: for bipartite graphs, instead of resorting to a coloring bound we can directly compute the size of a maximum (usually unfeasible) stable set in the current subgraph, resulting in much better bounds and shorter total running time.
- On the other hand, for general graphs we find that is better to spend less time on the upper bound computation: instead of recomputing a greedy coloring at each execution of Line 6, keeping the same coloring for several steps reduces the total running time.
- *Russian Doll Search*: to slightly restrict the number of visited nodes, we order the vertices as $v_1, \ldots, v_n$ by decreasing degree and call the algorithm n times: at step i, we include node i on our starting set S and restrict the set U to vertices v_j, with $j > i$, that are not neighbors of v_i.

4 Numerical Results

We now compare the performance of our formulation P_{arb} and our branch and bound algorithm on a benchmark of random graphs. We remark that the CVC problem is most interesting in graphs where the solution of CVC is strictly larger than the minimum vertex cover (we call such graphs *interesting*): if this is not the case one could just use the state of the art methods for finding the minimum vertex cover, and check that it induces a connected subgraph. This poses challenges to forming a benchmark of interesting graphs, as for instance the standard DIMACS benchmark [19] does not contain interesting graphs as far as we could check. Hence we resorted to sparse, random graphs. In particular, half of our graphs are Erdős–Rényi random graphs with density equal to 0.05; the others are bipartite random graphs, with density ranging from 0.1 to 0.5. We remark that bipartite graphs often seem to be interesting, which makes sense intuitively as each part of the bipartition forms a (possibly sub-optimal) vertex cover that is not connected: for instance, in the complete bipartite graph $K_{n,n}$, a minimum vertex cover has size n, while a minimum connected vertex cover has size $n + 1$. Moreover, as mentioned in the introduction, bipartite graphs are one of the simplest graph classes for which the VC problem is polynomial and CVC is NP-hard, which makes them good candidates for studying the differences between the two problems.

The graphs are produced with the functions fast_gnp_random_graph() and bipartite.random_graph() from the Networkx package [18], and the name of the graph indicates the random seed: for instance, G_i is the random graph on 100 vertices with density 0.05 created by seed i. Some of the seeds are missing since we only consider connected graphs. The experiments are run on a processor Intel Core i5-4590 (4 cores) clocked at 3.3 GHz with 4 GB RAM. Algorithm 1 is coded in Python, version 3.7, and Networkx functions articulation_points() and greedy_color() are used to perform Lines 10 and 6 respectively. We refer to [3] for the code for Algorithm 1 and for producing the formulation P_{arb}.

As for the implementation of formulation P_{arb}, it is also done in Python 3.7 and Gurobi 9.0.3 is used as MIP solver. Default parameters are used, and the results are averaged over three runs to account for the performance variability of the solver.

Table 1 indicates the results for random graphs, and Table 2 for bipartite graphs. Columns VC, CVC indicate the sizes of the minimum vertex cover and connected vertex cover respectively. The columns B&B t, B&B n indicate the running time (in seconds) and the number of nodes of Algorithm 1, and similarly for P_{arb} t and P_{arb} n.

It is evident from this comparison that solving the CVC problem with our formulation P_{arb} is much faster than with Algorithm 1, by a factor of one up to three order of magnitudes for some of the instances. Algorithm 1 does not finish in the time limit (one hour) for one of the bipartite graphs of density 0.2. Clearly, this might be partially due to the naive implementation of Algorithm 1, which is not optimized for speed: for instance, in Line 10 one does not have to recompute all cut vertices every time, but could restrict the computation to a single connected component of an appropriate subgraph of G. However, implementing this using the appropriate functions of Networkx actually further slows down the algorithm, as more infor-

Table 1 Results for random graphs of low density (0.05)

Name (seed)	$\|V\|$	$\|E\|$	VC	CVC	B&B t	B&B n	P_{arb} t	P_{arb} n
G_1	100	252	58	60	65.6	23138	0.2	1
G_2	100	247	55	56	6.4	435	0.1	1
G_3	100	232	56	57	12.7	1742	0.2	1
G_4	100	238	58	59	17.9	2296	0.4	191
G_7	100	257	56	59	21.1	2700	0.3	14
G_9	100	254	58	60	100.3	21846	0.2	1
G_{13}	100	260	58	59	56.2	18766	0.3	7
G_{16}	100	263	56	58	18.1	3620	0.2	1
G_{24}	100	234	58	58	11.2	1788	0.2	1
G_{25}	100	264	61	61	28.6	4789	0.5	158

Table 2 Results for random bipartite graphs. The density of each graph is written in its name, with the random seed in brackets

Name (seed)	$\|V\|$	$\|E\|$	VC	CVC	B&B t	B&B n	P_{arb} t	P_{arb} n
$G_{0.1}(1)$	100	255	49	54	11.1	6635	0.1	1
$G_{0.1}(4)$	100	242	50	57	863.4	818251	0.23	1
$G_{0.2}(0)$	100	483	50	57	1h+	2mln+	2.7	393
$G_{0.2}(1)$	100	497	50	56	1314.9	999252	2.3	338
$G_{0.3}(0)$	100	753	50	55	1137.1	723409	4.2	88
$G_{0.3}(1)$	100	753	50	55	1266.4	874949	4.3	166
$G_{0.4}(0)$	100	1007	50	54	354.5	210209	3	1
$G_{0.4}(1)$	100	977	50	53	69.6	39614	2.1	1
$G_{0.5}(0)$	100	1254	50	53	73.5	38685	3.9	1
$G_{0.5}(1)$	100	1231	50	53	50.0	26071	5.4	1

mation needs to be carried by each node. Hence, obtaining a faster version of the algorithm would require more advanced data structures and tools. But we believe this would not be enough to match the speed of P_{arb}: a major limit of the algorithm is that the bound used in the pruning phase (Line 6) is the same as for the classical vertex cover problem, i.e. does not take connectivity into account. Finding a better bound that is specific to the CVC problem is a non-trivial challenge, that we leave as an open problem. On the other hand, since Gurobi solves P_{arb} using a very small number of branching nodes, it would seem that the bound of the linear relaxation of P_{arb} is reasonably tight. This suggests the idea of taking the best of both worlds and integrating a bound based on P_{arb} into a combinatorial branch and bound algorithm.

5 Conclusion

The CVC problem brings together two of the most natural concepts in graph theory: stable sets and vertex covers on one hand, connectedness and spanning trees on the other. This paper approaches the problem from a modeling perspective, giving exact mixed-integer formulations for solving the problem, and compares them with a simple branch and bound algorithm. We believe that further work needs to be done in both directions: while we focused on modeling the connectivity requirement, better formulations could be found by using tighter formulations of the vertex cover problem; on the other hand, finding a faster branch and bound algorithm is fascinating challenge, as it is unclear how to tailor the branching and pruning steps to the CVC problem. We conclude by mentioning some extensions of CVC that could be of interest.

The Tree Cover problem [9] is closely related to the CVC problem: given a graph with non-negative weights on the edges and numbers k, w one asks to find a connected vertex cover of size at most k whose induced subgraph admits a spanning tree of weight at most w. It is easy to see that our formulations given in Sect. 2 can be adapted to model the Tree Cover problem, and exploring this further is an interesting research direction.

A natural generalization of the CVC problem considers hypergraphs instead of graphs [13]. We remark that deciding whether a hypergraph contains a spanning tree is NP-hard [1], hinting that the hypergraph version of CVC might be significantly harder than the graph version. However, we believe that our formulations can be extended to the hypergraph setting, and intend to investigate further in the future.

Finally, a different direction of research would be to generalize the connectivity constraint in the CVC problem to a matroid constraint, i.e. requiring that the edges of the subgraph induced by our vertex cover are full-rank sets of a given matroid. To the best of our knowledge, problems of this kind have not been studied before. Modelling such problems with mixed-integer formulations would be a promising line of inquiry, as there are several extended formulations for special matroid polytopes [2, 4, 7, 11].

References

1. Andersen, L.D., Fleischner, H.: The np-completeness of finding a-trails in eulerian graphs and of finding spanning trees in hypergraphs. Discrete Appl. Math. **59**(3), 203–214 (1995)
2. Aprile, M.: Extended formulations for matroid polytopes through randomized protocols. Oper. Res. Lett. **50**(2), 145–149 (2022)
3. Aprile, M.: Some code for solving the cvc problem (2022). https://github.com/manuel-aprile/CVC
4. Aprile, M., Conforti, M., Fiorini, S., Faenza, Y., Huynh, T., Macchia, M.: Slack matrices, k-products, and 2-level polytopes. Discrete Appl. Math. (2022). https://doi.org/10.1016/j.dam.2022.07.028
5. Aprile, M., Faenza, Y.: Extended formulations from communication protocols in output-efficient time. Math. Progr. **183**(1), 41–59 (2020)
6. Aprile, M., Faenza, Y., Fiorini, S., Huynh, T., Macchia, M.: Extension complexity of stable set polytopes of bipartite graphs. In: International Workshop on Graph-Theoretic Concepts in Computer Science. pp. 75–87. Springer (2017)
7. Aprile, M., Fiorini, S.: Regular matroids have polynomial extension complexity. Math. Oper. Res. **47**(1), 540–559 (2022)
8. Aprile, M., Fiorini, S., Huynh, T., Joret, G., Wood, D.R.: Smaller extended formulations for spanning tree polytopes in minor-closed classes and beyond. Electr. J. Comb. **28**(4), P4.47 (2021)
9. Arkin, E.M., Halldórsson, M.M., Hassin, R.: Approximating the tree and tour covers of a graph. Inf. Proc. Lett. **47**(6), 275–282 (1993)
10. Cardinal, J., Levy, E.: Connected vertex covers in dense graphs. Theor. Comput. Sci. **411**(26–28), 2581–2590 (2010)
11. Conforti, M., Kaibel, V., Walter, M., Weltge, S.: Subgraph polytopes and independence polytopes of count matroids. Oper. Res. Lett. **43**(5), 457–460 (2015)
12. Edmonds, J.: Matroids and the greedy algorithm. Math. progr. **1**(1), 127–136 (1971)

13. Escoffier, B., Gourvès, L., Monnot, J.: Complexity and approximation results for the connected vertex cover problem in graphs and hypergraphs. J. Discrete Algor. **8**(1), 36–49 (2010)
14. Fernau, H., Manlove, D.F.: Vertex and edge covers with clustering properties: complexity and algorithms. J. Discrete Algor. **7**(2), 149–167 (2009)
15. Fiorini, S., Huynh, T., Joret, G., Pashkovich, K.: Smaller extended formulations for the spanning tree polytope of bounded-genus graphs. Discrete & Comput. Geom. **57**(3), 757–761 (2017)
16. Garey, M.R., Johnson, D.S.: The rectilinear steiner tree problem is np-complete. SIAM J. Appl. Math. **32**(4), 826–834 (1977)
17. Guo, J., Niedermeier, R., Wernicke, S.: Parameterized complexity of vertex cover variants. Theory Comput. Syst. **41**(3), 501–520 (2007)
18. Hagberg, A., Swart, P., S Chult, D.: Exploring network structure, dynamics, and function using networkx. Technical Report, Los Alamos National Lab. (LANL), Los Alamos, NM (United States) (2008)
19. Johnson, D.S., Trick, M.A.: Cliques, coloring, and satisfiability: second DIMACS implementation challenge, October 11–13, 1993, vol. 26. American Mathematical Society (1996)
20. Kleinberg, J., Goemans, M.X.: The lovász theta function and a semidefinite programming relaxation of vertex cover. SIAM J. Discrete Math. **11**(2), 196–204 (1998)
21. Martin, R.K.: Using separation algorithms to generate mixed integer model reformulations. Oper. Res. Lett. **10**(3), 119–128 (1991)
22. Miller, C.E., Tucker, A.W., Zemlin, R.A.: Integer programming formulation of traveling salesman problems. J. ACM (JACM) **7**(4), 326–329 (1960)
23. Mölle, D., Richter, S., Rossmanith, P.: Enumerate and expand: improved algorithms for connected vertex cover and tree cover. Theory Comput. Syst. **43**(2), 234–253 (2008)
24. Padberg, M.W.: On the facial structure of set packing polyhedra. Math. Progr. **5**(1), 199–215 (1973)
25. Savage, C.: Depth-first search and the vertex cover problem. Inf. Proc. Lett. **14**(5), 233–235 (1982)
26. Wong, R.: Integer programming formulations of the traveling salesman problem. In: Proceedings of the 1980 IEEE International Conference on Circuits and Computers, pp. 149–152 (1980)
27. Wu, Q., Hao, J.K.: A review on algorithms for maximum clique problems. Eur. J. Oper. Res. **242**(3), 693–709 (2015)

Rigidity of Frameworks on Spheres

John Hewetson and Anthony Nixon

Abstract Consider the rigidity of bar-joint frameworks in 3-dimensional space that are constrained to lie on a union of spheres. It is well known that rigidity on a single sphere is equivalent to Euclidean rigidity and this equivalence extends to the case where the spheres are concentric. We consider the case when the spheres have distinct centres and give coloured sparsity conditions, analogous to the Euclidean case, necessary for a generic framework on the union of two spheres with different centres to be rigid. We show that these conditions are not sufficient in general and add additional conditions which we prove are sufficient in a special case.

1 Introduction

A bar-joint *framework* (G, p) is the combination of a finite, simple graph $G = (V, E)$ and a map $p : V \to \mathbb{R}^d$. The framework is *rigid* if the only edge-length preserving continuous deformations of the vertices arise from isometries of $\mathbb{R}^d$. In general it is NP-hard to determine the rigidity of a given framework [1], however for 'generic' frameworks one can linearise and consider the equivalent notion of infinitesimal rigidity [2]. Infinitesimal rigidity has been studied intensely in recent decades. Notably, when $d \leq 2$ there is a precise combinatorial characterisation [7, 12] which leads to efficient deterministic algorithms [8]. However, when $d \geq 3$ fundamental questions remain open [5, 14].

A natural question is what happens when $\mathbb{R}^d$ is replaced by a, perhaps smooth, d-dimensional manifold. The case of the unit sphere centred at the origin, $\mathbb{S}^d$, is well understood. In particular, due to work going back to Pogorelov [4, 11, 13], infinitesimal rigidity is now well understood as a projective invariant and hence infinitesimal rigidity in $\mathbb{R}^d$ is equivalent to infinitesimal rigidity on $\mathbb{S}^d$. This equivalence breaks

J. Hewetson · A. Nixon (✉)
Lancaster University, Mathematics and Statistics, Lancaster LA1 4YF, UK
e-mail: a.nixon@lancaster.ac.uk

J. Hewetson
e-mail: j.hewetson2@lancaster.ac.uk

A. Brieden et al. (eds.), *Graphs and Combinatorial Optimization: from Theory to Applications*, AIRO Springer Series 13,
https://doi.org/10.1007/978-3-031-46826-1_4

down for d-manifolds which 'support' isometry groups of smaller dimension. However, a number of recent papers have studied rigidity for 2-manifolds such as cylinders, surfaces of revolution, and surfaces arising from the more general context of linearly constrained frameworks [3, 9, 10]. Previous research has focused on either irreducible manifolds with a single connected component, or the mild extension to 'concentric' surfaces. One such example is concentric spheres, where the isometry group has the same dimension as any one of its irreducible components.

In this article we consider the reducible case proper by considering the rigidity of frameworks in $\mathbb{R}^3$ where the vertices lie on the union of two spheres with distinct centres. Since the dimension of the isometry group of a single sphere is different to that of the isometry group of such a union of spheres, the required number of constraints varies depending on whether a (sub-)framework is supported on one or both spheres. We use 2-(vertex)-coloured graphs to model such frameworks and analyse appropriate classes of 'coloured sparse' graphs. The multiple sparsity requirements lead to additional cases to check and substantial complications. One difficulty is the existence of various types of 'flexible circuit' (see Remark 1 below). It turns out that the problem is already non-trivial when exactly one vertex lives on the second sphere and our main result precisely characterises this case.

In Sect. 2 we introduce formal definitions, express infinitesimal rigidity as a matrix rank condition, and give coloured sparse graph counts necessary for infinitesimal rigidity. Then we show that certain operations on frameworks on a pair of non-concentric spheres preserve infinitesimal rigidity. Our main result is proved, via a recursive construction of the relevant coloured graphs, in Sect. 3.

2 Rigidity on Non-concentric Spheres

Let $\mathbb{S}^d_c$ denote the d-dimensional unit sphere centred at the point $c \in \mathbb{R}^d$. For notational convenience we will use $\mathbb{S}$ to denote the union $\mathbb{S}^2_{(0,0,0)} \cup \mathbb{S}^2_{(3,0,0)}$.[1] Throughout, unless stated otherwise, G will be a finite, simple graph with vertex set V, edge set E and $\chi : V \to \{r, b\}$ will be a 2-colouring of the vertices of G (with colours red and blue). The resulting ordered pair (G, χ) is a *2-coloured graph*, and we denote this object G_χ. For $i \in \{r, b\}$, let $V_i = \{v \in V : \chi(v) = i\}$. A *2-coloured subgraph* H_χ of G_χ is a subgraph H of G equipped with the colouring $\chi|_{V(H)}$. A bar-joint *framework* on $\mathbb{S}$, (G_χ, p), is the combination of a 2-coloured graph G_χ and a *realisation*, of G, $p : V \to \mathbb{R}^d$ such that $p(v) \in \mathbb{S}^2_{(0,0,0)}$ for all $v \in V_r$ and $p(v) \in \mathbb{S}^2_{(3,0,0)}$ for all $v \in V_b$. Throughout we will always assume that p is chosen in this way. That is, it will be assumed that χ and p are compatible in the sense that every $v \in V_r$ is mapped to $p(v) \in \mathbb{S}^2_{(0,0,0)}$ and every $u \in V_b$ is mapped to $p(u) \in \mathbb{S}^2_{(3,0,0)}$.

The frameworks (G_χ, p) and (G_χ, q) on $\mathbb{S}$ are: *equivalent* if $\|p(v_i) - p(v_j)\|^2 = \|q(v_i) - q(v_j)\|^2$ for all $v_i v_j \in E$; and *congruent* if $\|p(v_i) - p(v_j)\|^2 = \|q(v_i) -$

[1] Rigidity is invariant under both isometries and global dilations; we choose to work with unit radius spheres with the specified centres purely for convenience.

$q(v_j)\|^2$ for all $v_i, v_j \in V$. Note that (G, p) and (G, q) are congruent if and only if there exists an isometry ι of $\mathbb{R}^3$ such that $\iota(q) = p$. The framework (G_χ, p) on $\mathbb{S}$ is *$\mathbb{S}$-rigid* if there exists a neighbourhood of p such that every equivalent framework (G_χ, q) on $\mathbb{S}$, with q in that neighbourhood, is congruent to (G_χ, p). We refer to non-$\mathbb{S}$-rigid frameworks as *$\mathbb{S}$-flexible*. Moreover, (G_χ, p) is *minimally $\mathbb{S}$-rigid* if (G_χ, p) is $\mathbb{S}$-rigid and $((G - e)_\chi, p)$ is $\mathbb{S}$-flexible for any $e \in E$.

For a given framework on $\mathbb{S}$, (G_χ, p), let $\mathbb{Q}(p)$ denote the field extension of $\mathbb{Q}$ by adjoining the coordinates of (the image of) p. Then (G_χ, p) is *$\mathbb{S}$-generic* if $\mathrm{td}[\mathbb{Q}(p) : \mathbb{Q}] = 2|V|$. In other words the coordinates of p satisfy the polynomial equations defining the component of $\mathbb{S}$ the relevant point lives on but no other algebraic dependency. For generic frameworks it is standard to consider a linearisation through the Jacobian derivative matrix. We take the same approach and define the *rigidity matrix* for a framework on $\mathbb{S}$, (G_χ, p), to be the $(|E| + |V|) \times 3|V|$ matrix

$$\begin{pmatrix} R_3(G, p) \\ S(G_\chi, p) \end{pmatrix}$$

where $R_3(G, p)$ is the usual 3-dimensional rigidity matrix and $S(G_\chi, p)$ is a block diagonal matrix in which the 3-tuple in the row and columns for $v \in V$ is a normal vector to $\mathbb{S}$ at the point $p(v)$. We denote this matrix $R_\mathbb{S}(G_\chi, p)$. Borrowing language from matroid theory we say that a 2-coloured graph G_χ is *$\mathbb{S}$-independent* (resp. *$\mathbb{S}$-dependent*) if $R_\mathbb{S}(G_\chi, p)$ has linearly independent (resp. dependent) rows for some (and hence all) generic p. Moreover G_χ is an *$\mathbb{S}$-circuit* if $R_\mathbb{S}(G_\chi, p)$ has linearly dependent rows but deleting any single row of $R_\mathbb{S}(G_\chi, p)$ results in a matrix with linearly independent rows.

Let $G_\chi = (V, E)_\chi$ be a 2-coloured graph. When $|V_b|, |V_r| \geq 2$ then there is exactly one isometry that preserves $\mathbb{S}$; this congruence is the rotation about the line determined by the centres of the two spheres. However, if $V_b = \{x\}$ then a second isometry exists, namely rotation about the line determined by $p(x)$ and the centre of the 'red' sphere, and preserves congruence. Moreover, if $V_b = \emptyset$ then all three rotational isometries of the red sphere exist and preserve congruence.

Let $G_\chi = (V, E)_\chi$ be a 2-coloured graph, we may then define the function $f : \mathcal{P}(V) \setminus \emptyset \to \mathbb{Z}$ by

$$f(U) = \begin{cases} 3 & \text{if } U \subseteq V_r \text{ or } U \subseteq V_b, \\ 2 & \text{if } U \not\subseteq V_r, V_b \text{ and } |U \cap V_r| = 1 \text{ or } |U \cap V_b| = 1, \\ 1 & \text{if } |U \cap V_r|, |U \cap V_b| \geq 2. \end{cases}$$

We say that (G_χ, p) is *infinitesimally $\mathbb{S}$-rigid* if $\mathrm{rank}\, R_\mathbb{S}(G_\chi, p) = 3|V| - f(V)$ or $G \cong K_1$ and *minimally infinitesimally $\mathbb{S}$-rigid* if it is infinitesimally $\mathbb{S}$-rigid and the rows of $R_\mathbb{S}(G_\chi, p)$ are linearly independent. It follows immediately from [12] that when $f(V) = 3$ this rank can be achieved. It is not difficult to construct small examples where the other maximum possible ranks can be achieved. Figure 1 illus-

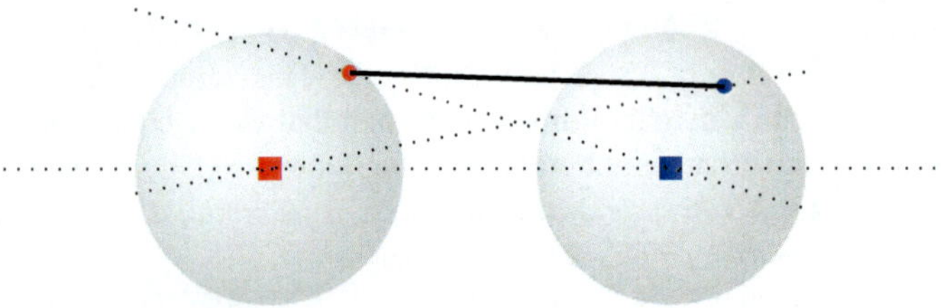

Fig. 1 A realisation of K_2 on two spheres. The three dotted lines represent axes of three rotational motions that result in congruent frameworks. This framework is $\mathbb{S}$-rigid even though it is not infinitesimally $\mathbb{S}$-rigid

trates how the number of vertices of each colour affects the number isometries of $\mathbb{R}^3$ that preserve $\mathbb{S}$, and hence why the count $f(U)$ depends on χ.

For $\mathbb{S}$-generic frameworks one may use the inverse function theorem to deduce the following analogue of a key theorem of Asimow and Roth [2] (see also [9] for the irreducible manifold case).

Proposition 1 *Let (G_χ, p) be an $\mathbb{S}$-generic framework on at least 5 vertices. Then (G_χ, p) is $\mathbb{S}$-rigid if and only if it is infinitesimally $\mathbb{S}$-rigid.*

Since the rank of the rigidity matrix is maximised at any $\mathbb{S}$-generic point, $\mathbb{S}$-rigidity depends only on the underlying 2-coloured graph and hence we say that G_χ is $\mathbb{S}$-rigid if some (and hence any) $\mathbb{S}$-generic framework (G_χ, p) on $\mathbb{S}$ is $\mathbb{S}$-rigid.

To see the necessity of the lower bound on the number of vertices, consider K_2 (see Fig. 1), K_3, and K_4. Regardless of the colouring assigned by χ, any complete 2-coloured graph is $\mathbb{S}$-rigid. However: $(K_2)_\chi$ is infinitesimally $\mathbb{S}$-rigid if and only if χ is monochrome; $(K_3)_\chi$ is infinitesimally $\mathbb{S}$-rigid if and only if χ assigns all the vertices of K_3 the same colour; and $(K_4)_\chi$ is infinitesimally $\mathbb{S}$-rigid if and only if one colour is assigned to at most one vertex.

Let $G = (V, E)$ be a graph and take $\emptyset \neq X \subseteq V$. Let $i_G(X)$, or $i(X)$ if the graph is clear from the context, denote the number of edges in the subgraph of G induced by X. A 2-coloured graph G_χ is $(2, f)$-*sparse* if for all $X \subseteq V$ such that $|X| \geq 2$, $i_G(X) \leq 2|X| - f(X)$. G_χ is $(2, f)$-*tight* if it is $(2, f)$-sparse and $|E| = 2|V| - f(V)$.

Proposition 2 *Let G_χ be minimally infinitesimally $\mathbb{S}$-rigid on at least 2 vertices. Then G_χ is $(2, f)$-tight.*

Proof As G_χ is infinitesimally $\mathbb{S}$-rigid, $|E| \geq 2|V| - f(V)$. Suppose $i(X) > 2|X| - f(X)$ for some $X \subseteq V$ with $|X| \geq 2$. Since rank $R_\mathbb{S}(G_\chi[X], p|_X) \leq 3|X| - f(X)$, $R_\mathbb{S}(G_\chi[X], p|_X)$ has linearly dependent rows, contradicting the fact that G_χ is minimally infinitesimally $\mathbb{S}$-rigid. Thus G_χ is $(2, f)$-sparse and hence $(2, f)$-tight. □

The remainder of this article will consider the converse question. That is, given a $(2, f)$-tight graph is it $\mathbb{S}$-rigid? We first illustrate one reason this is challenging. It is immediate from Proposition 2 that $\mathbb{S}$-circuits admit different behaviour to standard

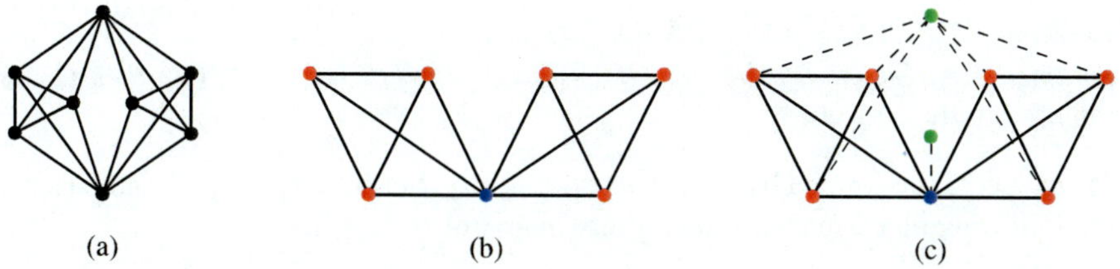

Fig. 2 **a** the double banana graph which satisfies the Maxwell conditions for 3-dimensional rigidity but is not rigid. **b** A 2-coloured graph that is $(2, f)$-tight but not $\mathbb{S}$-rigid. **c** An augmentation of (**a**) that models the constraints under the two spheres models of the coloured graph in (**b**). The green vertices represent the fixed centres of the two spheres and the dashed edges to the green vertices represent the constraints that the red/blue vertices must move on the given sphere

Fig. 3 Illustration of the 0- and 1-extension operations

rigidity models. Specifically a *flexible* $\mathbb{S}$*-circuit* is a graph that is both $\mathbb{S}$-flexible and an $\mathbb{S}$-circuit. Somewhat surprisingly, flexible $\mathbb{S}$-circuits exist. For contrast, in the well studied 2-dimensional Euclidean rigidity (or equivalently for rigidity on the 2-sphere) all circuits are rigid. The following example shows the well known double banana graph (see Fig. 2) can easily be disguised as a flexible $\mathbb{S}$-circuit.

Remark 1 Let G_χ be the coloured graph consisting of two copies of K_4 that intersect in exactly one vertex, which is the unique blue vertex of G_χ (see Fig. 2b). We may consider the corresponding *'linearly constrained'* framework (see [3]), depicted in Fig. 2c, which has two additional vertices located at the centers of the two spheres. This framework clearly contains the 'usual' double banana (Fig. 2a) and admits the same infinitesimal motion. Moreover, it is easily checked that G_χ is $(2, f)$-sparse, with $f(V) = 2$, and hence the converse to Proposition 2 is false. Further note that one may adapt the example by careful replacement of one, or both, of the K_4 subgraphs with larger graphs to create infinite families of such examples in both the $f(V) = 2$ and $f(V) = 1$ cases.

The existence of flexible $\mathbb{S}$-circuits makes characterising $\mathbb{S}$-rigidity challenging. For that reason we focus on, and resolve, the case where $f(V) = 2$. Note that this case is rich enough to contain the flexible $\mathbb{S}$-circuits described in Remark 1.

We conclude this section by deriving two ways to build $\mathbb{S}$-rigid coloured graphs. To that end, G' is said to be obtained from a graph G by a *0-extension* if $G = G' - v$ for a vertex $v \in V(G')$ of degree 2, or by a *1-extension* if $G = G' - v + xy$ for a vertex $v \in V(G')$ of degree 3 where x and y are neighbours of v. These operations are illustrated in Fig. 3. The converse operations will be referred to as *0-reduction* and *1-reduction* respectively.

Lemma 1 *Let G_χ be infinitesimally $\mathbb{S}$-rigid on at least 2 vertices and let G' be obtained from G by a 0-extension. If $\chi'|_V = \chi$ and $f(V(G')) = f(V)$ then $G'_{\chi'}$ is infinitesimally $\mathbb{S}$-rigid.*

Proof Let G' be formed by adding v and vv_1, vv_2. Set $p' = (p, p'(v))$ and observe the following block form for the rigidity matrix of $(G'_{\chi'}, p')$:

$$R_{\mathbb{S}}(G'_{\chi'}, p') = \begin{pmatrix} R_{\mathbb{S}}(G_\chi, p) & 0 \\ * & T \end{pmatrix}.$$

Here T is a 3×3 matrix whose entries are determined by p'. In particular, $(G'_{\chi'}, p')$ is infinitesimally $\mathbb{S}$-rigid if and only if T is invertible. It is easy to check this holds for any non-collinear triple $p(v_1), p(v_2), p(v)$ and hence it holds for $\mathbb{S}$-generic p'. □

Lemma 2 *Let G_χ be infinitesimally $\mathbb{S}$-rigid on at least 2 vertices and let G' be obtained from G by a 1-extension that deletes the edge v_1v_2 and adds a new vertex v with $N(v) = \{v_1, v_2, v_3\}$. If $\chi'|_V = \chi$, $f(V(G')) = f(V)$ and $v, v_1 \in V_r$ then $G'_{\chi'}$ is infinitesimally $\mathbb{S}$-rigid.*

Proof If $v_2 \in V_r$ then we place $p'(v)$ on the great circle of $\mathbb{S}_{(0,0,0)}$ through the points $p(v_1)$ and $p(v_2)$. It is easy to check that the complete graph on v_1, v_2, v is a minimally dependent set. If $v_2 \in V_b$ then we place $p(v)$ on the intersection of $\mathbb{S}_{(0,0,0)}$ and the unique line determined by $p(v_1)$ and $p(v_2)$ (generically this intersection necessarily contains a point distinct from $p(v_1)$). It is again easy to check that the complete graph on v_1, v_2, v is a minimally dependent set. Hence in both cases we may use the fact that 0-extension preserves infinitesimal $\mathbb{S}$-rigidity when $p(v), p(v_1), p(v_3)$ are not collinear to show that $(G + v + \{v_1v, v_3v\})_{\chi'}$ is infinitesimally $\mathbb{S}$-rigid. Now by the minimality of the dependent set on v_1, v_2, v we may remove the edge v_1v_2 and add the edge v_2v without altering the rank of $R_{\mathbb{S}}(G_\chi, p)$. This shows that $(G'_{\chi'}, p')$, and hence $G'_{\chi'}$, is infinitesimally $\mathbb{S}$-rigid. □

3 Rigidity of Nearly Monochrome Graphs

A 2-coloured graph G_χ is *monochrome* if every vertex has the same colour and *nearly monochrome* if it is not monochrome but there exists $v \in V$ such that $(G - v)_\chi$ is monochrome. The classical characterisation of rigidity on the 2-sphere [12, 13] resolves the case when G_χ is monochrome. In this section we consider the case when G_χ is nearly monochrome and hence one of the spheres contains exactly one vertex of the graph. In this case the combinatorial conditions simplify to $i_G(X) \leq 2|X| - 2$ for all $X \subseteq V$, and $i_G(X) \leq 2|X| - 3$ for all $X \subseteq V_r$ with $|V_r| > 1$.

For a graph G and $v \in V$ we will use $d_G(v)$, $N_G(v)$, and $N_G[v]$ to denote the degree, neighbourhood, and closed neighbourhood respectively of v in G. Furthermore, for $X, Y \subseteq V$ we use $d_G(X, Y)$ to denote the number of edges of the form xy

such that $x \in X \setminus Y$ and $y \in Y \setminus X$. In each case we may drop the subscript if the graph is clear from the context. We will repeatedly use the equality

$$i_G(X) + i_G(Y) + d_G(X, Y) = i_G(X \cup Y) + i_G(X \cap Y). \quad (1)$$

Given a 2-coloured graph G_χ, we say that $\emptyset \neq X \subseteq V$ is *f-critical* (in G_χ) if $i_G(X) = 2|X| - f(X)$. If G_χ is $(2, f)$-sparse then a 1-reduction of G_χ is said to be *admissible* if the resulting 2-coloured graph is $(2, f)$-sparse. The next lemma shows that f-critical sets are the main 'blockers' to admissible 1-reductions in $(2, f)$-tight graphs. Vertices of degree 3 shall be referred to as *nodes*. We say that a node $v \in V_r$ with $N(v) = \{x, y, z\}$ is: *type 1* if $N(v) \subseteq V_r$ and *type 2* if $|N(v) \cap V_b| = 1$. The other type of node possible in a nearly monochrome graph is a node $v \in V_b$ with $N(v) \subseteq V_r$; we avoid working with such nodes.

The proof of the next lemma is standard and hence omitted. For the proof and additional details on subsequent results in this section see [6, Chap. 5].

Lemma 3 *Let G_χ be a 2-coloured graph, suppose v is a node of G and let $N(v) = \{x, y, z\}$. If G_χ is $(2, f)$-sparse then $(G - v + xy, \chi|_{V \setminus \{v\}})$ is not a $(2, f)$-sparse 2-coloured graph if and only if $xy \in E$ or there exists an f-critical set $Z \subseteq V$ such that $\{x, y\} \subseteq Z$. Further, if such a set Z exists and v is type 1 or type 2 then $z \notin Z$.*

Given that G_χ is $(2, f)$-sparse and v is a node of G, v is *admissible* if at least one of the 1-reductions of G_χ at v result in a $(2, f)$-sparse 2-coloured graph. To understand this, the remainder of this subsection consider interactions between f-critical sets.

Lemma 4 *Let G_χ be $(2, f)$-sparse and take $\emptyset \neq X, Y \subseteq V$. If X and Y are f-critical in G_χ and $|X \cap Y| \geq 2$ then $d(X, Y) = 0$, $X \cup Y$ is f-critical in G_χ and either*

1. *$f(X \cup Y) = 1$, $f(X) = 2 = f(Y)$, $f(X \cap Y) = 3$, and $X \cap Y$ is f-critical in G_χ, or*
2. *$f(X \cup Y) = 1$, $f(X) = 2 = f(Y) = f(X \cap Y)$, and $G[X \cap Y] \cong K_2$, or*
3. *$f(X \cup Y) = \min\{f(X), f(Y)\}$, $f(X \cap Y)\} = \max\{f(X), f(Y)\}$, and $X \cap Y$ is f-critical in G_χ.*

Proof As G_χ is $(2, f)$-sparse, and X and Y are f-critical in G_χ, it follows from Eq. (1) that $f(X) + f(Y) \geq f(X \cup Y) + f(X \cap Y) + d(X, Y)$. Note that, by the definition of f,

$$f(X \cap Y) \geq \max\{f(X), f(Y)\} \geq \min\{f(X), f(Y)\} \geq f(X \cup Y). \quad (2)$$

Let $D = \min\{f(X), f(Y)\} - f(X \cup Y)$. If $D = 2$ then $f(X \cup Y) = 1$ and $f(X) = 3 = f(Y)$. However as $X \cap Y \neq \emptyset$ this contradicts the definition of f. Hence $D \in \{0, 1\}$.

Suppose instead that $D = 1$, so $f(X \cup Y) \leq 2$. If $f(X \cup Y) = 2$ then $f(X) = 3 = f(Y)$ and, as in the previous paragraph, we have a contradiction. Therefore $f(X \cup Y) = 1$ and $\min\{f(X), f(Y)\} = 2$. If $\max\{f(X), f(Y)\} = 3$ then, as $|X \cap$

$Y| \geq 2$ and $\min\{f(X), f(Y)\} = 2$, it follows that $f(X \cup Y) = 2$, a contradiction. Therefore $f(X) = 2 = f(Y)$, and so $f(X \cap Y) \in \{2, 3\}$. If $f(X \cap Y) = 3$ then it follows that $d(X, Y) = 0$ and then Eq. (1) implies that $X \cup Y$ and $X \cap Y$ are both f-critical in G_χ. Alternatively, if $f(X \cap Y) = 2$ then, as $f(X \cup Y) = 1$, $|X \cap Y| = 2$. It follows from Eq. (1) that $d(X, Y) = 0$ and so $i(X \cup Y) = 2|X \cup Y| - 1$ and $i(X \cap Y) = 2|X \cap Y| - 3$. That is, $X \cup Y$ is f-critical in G_χ and $G[X \cap Y] \cong K_2$.

Finally suppose that $D = 0$, so $f(X \cup Y) = \min\{f(X), f(Y)\}$. It now follows from Eq. (2) that $d(X, Y) = 0$ and $f(X \cap Y) = \max\{f(X), f(Y)\}$. Consequently, Eq. (1) implies that both $X \cup Y$ and $X \cap Y$ are f-critical in G_χ. □

Lemma 5 *Let G_χ be $(2, f)$-sparse and take $\emptyset \neq X, Y \subseteq V$. If X and Y are f-critical in G_χ, $|X \cap Y| = 1$, $f(V) \geq 2$ and $d(X, Y) \geq 1$ then $f(X \cup Y) = \min\{f(X), f(Y)\}$, $d(X, Y) = 1$, $X \cup Y$ is f-critical in G_χ, and $\max\{f(X), f(Y)\} = 3$.*

Proof As G_χ is $(2, f)$-sparse, and X and Y are f-critical in G_χ and $|X \cap Y| = 1 \leq d(X, Y)$ it follows from Eq. (1) that

$$f(X) + f(Y) \geq f(X \cup Y) + f(X \cap Y) + d(X, Y) \geq f(X \cup Y) + 2 + 1. \quad (3)$$

Let $D = \min\{f(X), f(Y)\} - f(X \cup Y)$. If $D = 1$ then, as $f(V) \geq 2$, $f(X \cup Y) = 2$ and $f(X) = f(Y) = 3$. However, as $X \cap Y \neq \emptyset$ this contradicts the definition of f. Hence $D = 0$, so $f(X \cup Y) = \min\{f(X), f(Y)\}$. It now follows from Eq. (3) that $d(X, Y) = 1$ and $\max\{f(X), f(Y)\} = 3$. Consequently, Eq. (1) implies that $X \cup Y$ is f-critical in G_χ. □

Similar considerations give the following result, see [6, Chap. 5] for the proof.

Lemma 6 *Let G_χ be $(2, f)$-sparse and take $\emptyset \neq X, Y, Z \subseteq V$. If X, Y and Z are f-critical in G_χ, $|X \cap Y| = |X \cap Z| = |Y \cap Z| = 1$, $X \cap Y \cap Z = \emptyset$ and $f(V) \geq 2$ then $f(X) + f(Y) + f(Z) \geq 8$, $f(X \cup Y \cup Z) = \min\{f(X), f(Y), f(Z)\}$, $X \cup Y \cup Z$ is f-critical in G_χ, $d(X \cup Y, Z) = 0$, and $d(X \cap Z, Y \cap Z) = d(X, Y)$.*

Lemma 7 *Let G_χ be $(2, f)$-sparse, suppose $v \in V$ is a node, and let $N(v) = \{x, y, z\}$. If $f(V) \geq 2$, and v is type 1 or type 2, then v is non-admissible if and only if v is type 2 and either (i) $G[N(v)] \cong K_3$, or (ii) $G[N(v)] = (N(v), \{xz, yz\})$, and there exists $Z \subseteq V$ such that $N[v] \cap Z = \{x, y\}$, $f(Z) = 3$, and Z is f-critical in G_χ.*

Proof If v is type 2 and (i) or (ii) hold then Lemma 3 implies that v is non-admissible. On the other hand, let us suppose that v is non-admissible. Let $H = (N(v), \{xy, xz, yz\})$ and let $F = E(H) \setminus E$. We deal with both types of v simultaneously, and proceed by considering $i(N(v))$.

If $i(N(v)) = 0$ then let $F = \{e_1, e_2, e_3\}$. As v is non-admissible, Lemma 3 implies there exist $U_1, U_2, U_3 \subseteq V$ such that, for $i \in \{1, 2, 3\}$, the endpoints of e_i are in U_i and U_i is f-critical in G_χ. As $f(V) \geq 2$, Lemmas 4 and 6 together imply that there exist $i, j \in \{1, 2, 3\}$ such that $i \neq j$ and $U_i \cup U_j$ is f-critical in G_χ, or $U_1 \cup U_2 \cup U_3$

is f-critical in G_χ. However, as v is type 1 or type 2 this contradicts Lemma 3. If $i(N(v)) = 1$ then let $F = \{e_1, e_2\}$. As v is non-admissible, Lemma 3 implies there exist $U_1, U_2 \subseteq V$ such that, for $i \in \{1, 2\}$, the endpoints of e_i are in U_i and U_i is f-critical in G_χ. As $d(U_1, U_2) \geq 1$ and $f(V) \geq 2$, Lemma 4 and Lemma 5 together imply that $U_1 \cup U_2$ is f-critical in G_χ. However, as v is type 1 or type 2 and G_χ is $(2, f)$-sparse this contradicts Lemma 3.If $i(N(v)) = 2$ then let $F = \{e\}$. As v is non-admissible, Lemma 3 implies there exists $U \subseteq V$ such that the endpoints of e are in U and U is f-critical in G_χ. As G_χ is $(2, f)$-sparse and v is type 1 or type 2, Lemma 3 implies $|U \cap N(v)| = 2$. Consequently,

$$i(U \cup N[v]) \geq i(U) + 5 = (2|U| - f(U)) + 5 = 2|U \cup N[v]| - (f(U) - 1),$$

As G_χ is $(2, f)$-sparse it follows that $f(U) \geq f(U \cup N[v]) + 1$. So, as $f(V) \geq 2$, it follows that $f(U) = 3$ and $f(U \cup N[v]) = 2$. Therefore v is type 2 and (ii) holds. If $i(N(v)) = 3$ then, as G_χ is $(2, f)$-sparse, v is type 2 and (i) holds. □

A nearly monochrome graph G_χ is $(2, f)$-*cut-sparse* if it is $(2, f)$-sparse and every $(2, f)$-tight subgraph of G_χ with at least three vertices is 2-connected. G_χ is $(2, f)$-*cut-tight* if it is $(2, f)$-cut-sparse and $(2, f)$-tight.

Lemma 8 *Let G_χ be nearly monochrome and $(2, f)$-cut-sparse. Suppose $v \in V$ is a node and $N(v) = \{x, y, z\}$. If the 1-reduction of G_χ at v adding xy is admissible then it is non-feasible if and only if there exist $W_1, W_2 \subseteq V$ such that $N(v) \cap W_1 = \{x, y\}$, $W_1 \cap W_2 = \{v \in V : f(V \setminus \{v\}) > f(V)\}$, $i(W_1) = 2|W_1| - 3$, W_2 is f-critical in G_χ, and $d(W_1, W_2) = 0$.*

Proof Let $G'_{\chi'}$ denote the 2-coloured graph resulting from the 1-reduction of G_χ at v adding xy. Suppose there exist such sets W_1 and W_2. Then W_1 and W_2 are f-critical in $G'_{\chi'}$. As $G'_{\chi'}$ is $(2, f)$-sparse, Eq. 1 implies $W_1 \cup W_2$ is f-critical in $G'_{\chi'}$. So $G'[W_1 \cup W_2]_{\chi'}$ is a $(2, f)$-tight 2-coloured subgraph of $G'_{\chi'}$ and $|W_1 \cup W_2| \geq 3$. However it is not 2-connected and therefore the 1-reduction is not feasible.

Conversely, suppose the 1-reduction is not feasible. Then there exists $U \subseteq V(G')$ such that $|U| \geq 3$, U is f-critical in $G'_{\chi'}$ and there exists a cut-vertex, u, of $G'[U]$. As U is f-critical in $G'_{\chi'}$ and u is a cut-vertex of $G'[U]$ we note that $f(U) = 2$. Let $H_1, \ldots, H_n$ denote the components of $G'[U \setminus \{u\}]$. It follows from Eq. 1 that for all $1 \leq i \leq n$, $V(H_i) \cup \{u\}$ is f-critical in $G'_{\chi'}$ and $f(V(H_i) \cup \{u\}) = 2$. Moreover, as $f(V) = 2$, $\{u\} = \{w \in V : f(V \setminus \{w\}) > f(V)\}$. As G_χ is $(2, f)$-cut-sparse, there exists $1 \leq i \leq n$ such that $\{x, y\} \subseteq V(H_i) \cup \{u\}$. We may suppose that $\{x, y\} \subseteq V(H_1) \cup \{u\}$. Then $(V(H_1) \cup \{u\}) \cap (V(H_2) \cup \{u\}) = \{u\}$, $i_G(V(H_1) \cup \{u\}) = 2|V(H_1) \cup \{u\}| - 3$, $V(H_2) \cup \{u\}$ is $(2, f)$-critical in G_χ and $d_G(V(H_1) \cup \{u\}, V(H_2) \cup \{u\}) = 0$. □

We next show that, at every admissible vertex, there exists some feasible reduction. See Fig. 4 for an example.

Lemma 9 *Let G_χ be nearly monochrome and $(2, f)$-cut-sparse and suppose $v \in V$ is a node. Then v is feasible if and only if v is admissible.*

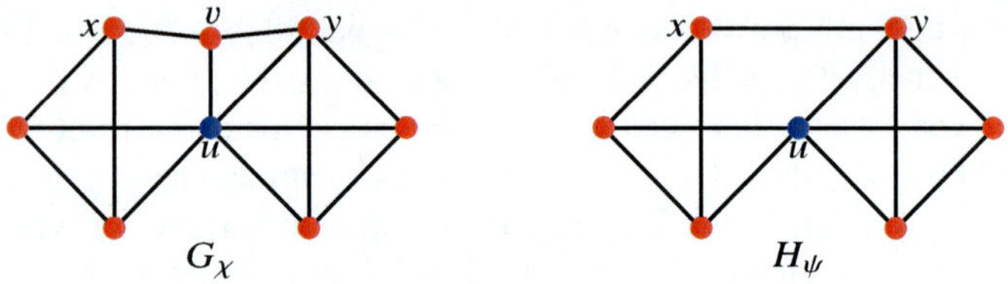

Fig. 4 The three possible 1-reductions of G_χ at v demonstrate distinct outcomes. The 1-reduction adding uy is non-admissible, the 1-reduction adding ux is admissible but non-feasible, and the 1-reduction adding xy (resulting in H_ψ) is feasible

Proof If v is feasible then clearly v is admissible. On the other hand, let us suppose that v is admissible. Let $V_b = \{u\}$. Let $H = (N(v), \{xy, xz, yz\})$ and let $F = E(H) \setminus E$. We present only the case when $i(N(v)) = 1$. The other cases use similar analysis and can be found in [6, Chap. 5]. Let $F = \{e_1, e_2\}$. As v is admissible we may suppose, without loss of generality, that the 1-reduction of G_χ at v adding e_1 is admissible. Suppose there exist $W_1, W_2 \subseteq V$ such that $N(v) \cap W_1 = \{\text{endpoints of } e_1\}$, $W_1 \cap W_2 = \{u\}$, $i(W_1) = 2|W_1| - 3$, W_2 is f-critical in G_χ and $d(W_1, W_2) = 0$. There are two cases. Firstly, suppose there exists $U \subseteq V$ such that $U \cap N[v] = \{\text{endpoints of } e_2\}$ and U is f-critical in G_χ. Then Eq. 1 implies

$$i(W_1 \cup U) \geq 2|W_1 \cup U| + 2|W_1 \cap U| - (2 + f(U) + i(W_1 \cap U)).$$

Note that, as $W_1 \cap U \neq \emptyset$ and $f(V) = 2 = f(W_1)$, $f(U) = f(W_1 \cap U)$. If $|W_1 \cap U| \geq 2$ then, as G_χ is $(2, f)$-sparse, it follows that

$$i(W_1 \cup U) \geq (2|W_1 \cup U| - 2) + (2|W_1 \cap U| - f(W_1 \cap U)) - i(W_1 \cap U) \geq 2|W_1 \cup U| - 2$$

and so $W_1 \cup U$ is $(2, f)$-critical in G_χ. Then $i(W_1 \cup U \cup \{v\}) = 2|W_1 \cup U \cup \{v\}| - 1$, a contradiction. Hence $|W_1 \cap U| = 1$. Thus $i(W_1 \cup U) \geq 2|W_1 \cup U| - f(U)$. Hence $i(W_1 \cup U \cup \{v\}) \geq 2|W_1 \cup U \cup \{v\}| - (f(U) - 1)$. As $f(V) \geq 2$ it follows that $W_1 \cup U \cup \{v\}$ is f-critical in G_χ, $f(U) = 3$, $i(W_1 \cup U) = 2|W_1 \cup U| - f(U)$, and $d(W_1, U) = 1$. If $W_2 \cap U \neq \emptyset$ then Lemma 4 implies $(W_1 \cup U \cup \{v\}) \cap W_2$ is f-critical in G_χ. This implies $d_{G[(W_1 \cup U \cup \{v\}) \cap W_2]}(w) \geq 2$, a contradiction. If $W_2 \cap U = \emptyset$, Eq. 1 implies $(W_1 \cup U \cup \{v\}) \cup W_2$ is f-critical in G_χ and $d(W_1 \cup U \cup \{v\}, W_2) = 0$. So u is a cut-vertex of $G[W_1 \cup U \cup \{v\} \cup W_2]$, a contradiction.

Alternatively, suppose there exist $U_1, U_2 \subseteq V$ such that $U_1 \cap U_2 = \{u\}$, $N(v) \cap U_1 = \{\text{endpoints of } e_2\}$, $i(U_1) = 2|U_1| - 3$, U_2 is f-critical in G_χ and $d(U_1, U_2) = 0$. Then Eq. 1 implies that $i(W_1 \cup U_1) \geq (2|W_1| - 3) + (2|U_1| - 3) + 1 - i(W_1 \cap U_1) = 2|W_1 \cup U_1| + 2|W_1 \cap U_1| - (5 + i(W_1 \cap U_1))$. As G_χ is $(2, f)$-sparse, it follows that $i(W_1 \cup U_1) \geq (2|W_1 \cup U_1| - 3) + (2|W_1 \cap U_1| - 2) - i(W_1 \cap U_1) \geq 2|W_1 \cup U_1| - 3$. Hence $i(W_1 \cup U_1 \cup \{v\}) \geq 2|W_1 \cup U_1 \cup \{v\}| - 2$. It follows that $W_1 \cup U_1 \cup \{v\}$ is f-critical in G_χ, $i(W_1 \cup U_1) = 2|W_1 \cup U| - 3$, and $d(W_1, U) = 1$. As $|W_2 \cap U_2| \geq 1$ and G_χ is $(2, f)$-cut-sparse, $|W_2 \cap U_2| \geq 2$. Therefore Lemma 4 implies $W_2 \cap U_2$ is f-critical in G_χ. As $(W_2 \cap U_2) \cap (W_1 \cup U_1 \cup \{v\}) = \{u\}$, Eq. 1 implies $(W_1 \cup U_1 \cup \{v\}) \cup (W_2 \cap U_2)$ is f-critical in G_χ. However, then u is a cut-vertex of $G[(W_1 \cup U_1 \cup \{v\}) \cup (W_2 \cap U_2)]$, a contradiction. □

Theorem 1 *A nearly monochrome graph G_χ on at least 2 vertices is minimally infinitesimally $\mathbb{S}$-rigid if and only if G_χ is $(2, f)$-cut-tight.*

Proof The necessity of the $(2, f)$-tight count with $f(V) = 2$ was proved in Proposition 2. Suppose G_χ is minimally infinitesimally $\mathbb{S}$-rigid and has a $(2, f)$-tight subgraph H_χ, with $|V(H)| \geq 3$, which is not 2-connected. Then $f(V(H)) = 2$ and H is the union of two $(2, f)$-tight subgraphs intersecting in a single vertex u. It follows that H_χ is $\mathbb{S}$-dependent, as in Remark 1, a contradiction.

For the sufficiency we proceed via induction on $|V|$. The base case is any nearly monochrome $(K_4)_\chi$. It is not hard to show that $R_{\mathbb{S}}((K_4)_\chi, p)$ has full rank for any $\mathbb{S}$-generic p. So we may suppose $G_\chi \neq (K_4)_\chi$. Suppose G_χ is $(2, f)$-cut-tight. There exists $v \in V$ of minimum degree with $f(V \setminus \{v\}) = f(V)$. Note that $d(v) \in \{2, 3\}$. $(G - v)_\chi$ is a nearly monochrome subgraph of G_χ and hence is $(2, f)$-cut-sparse. If $d(v) = 2$ then $i(V \setminus \{v\}) = i(V) - 2 = 2|V \setminus \{v\}| - 2$, so $(G - v)_\chi$ is $(2, f)$-cut-tight and we can complete the proof using Lemma 1.

Alternatively suppose $d(v) = 3$, and, for $i \in \{1, 2\}$, let $W_i = \{w \in V : w \text{ is a type } i \text{ node of } G_\chi\}$. Let $V_b = \{u\}$. Then

$$4|V| - 4 = 2|E| \geq 3(|W_1| + |W_2|) + d(u) + 4(|V| - (|W_1| + |W_2| + 1)).$$

Hence $|W_1| + |W_2| \geq d(u) \geq |W_2|$. If $W_1 = \emptyset$ then $d(u) = |W_2|$ and so Lemma 7 implies that for all $w \in W_2$, w is admissible or $G[N(w)] \cong K_4$. In the latter case, as $G \ncong K_4$ and G_χ is $(2, f)$-cut-tight, this leads to a contradiction. Hence $W_1 \neq \emptyset$ and so, by Lemma 7, there exists an admissible node of G_χ. Now Lemma 9 implies there exists a feasible node of G_χ and so we can complete the proof using Lemma 2. □

4 Concluding remarks

Well known graph orientation type algorithms (e.g. [8]) can be easily adapted to prove that we can determine whether a graph is $(2, f)$-cut-tight in polynomial time and hence determine generic rigidity for nearly monochrome graphs.

Motivated by Theorem 1 we provide an additional necessary condition for $\mathbb{S}$-rigidity in the more general case. To do this we augment the definition of $(2, f)$-cut-tight. Given a $(2, f)$-tight graph G_χ we say that G_χ is $(2, f)$*-cut-tight* if any cut-vertex v (if any exist) of any $(2, f)$-tight subgraph H_χ of G_χ gives rise to components of $H - v$ that are all not monochromatic.

Proposition 3 *If G_χ is minimally infinitesimally $\mathbb{S}$-rigid on at least 2 vertices then G_χ is $(2, f)$-cut-tight.*

Proof By Proposition 2, G_χ is $(2, f)$-tight. Suppose H_χ is a $(2, f)$-tight subgraph of G_χ, with $|V(H)| \geq 3$, and there exists a cut-vertex, v, of H such that $H - v$ has a monochrome component. Then $f(V(H)) \in \{1, 2\}$ and H is the union of (at least) two $(2, f)$-tight subgraphs intersecting in the single vertex v. Since H_χ is $(2, f)$-tight, v is coloured differently from the monochrome component of $H - v$. It follows that H_χ is $\mathbb{S}$-dependent, as in Remark 1, a contradiction. □

It is tempting to conjecture that every $(2, f)$-cut-tight graph is $\mathbb{S}$-rigid. In the remaining case, when $f(V) = 1$, Lemmas 1 and 2 can be used to verify that statement for a large family of $(2, f)$-cut-tight graphs. However a full resolution seems beyond presently available techniques. The conjecture would be resolved by characterising $\mathbb{S}$-independence for an arbitrary 2-coloured graph G_χ. We conclude the paper by noting that combining the recursive construction of [7] with Lemma 1 and a stronger version of Lemma 2 shows that if G is $(2, 3)$-sparse then G_χ is $\mathbb{S}$-independent (for any χ). This shows that the difficult part of the conjecture is dealing with 'dense' coloured subgraphs.

Acknowledgements A.N. was partially supported by EPSRC grant number EP/W019698/1.

References

1. Abbott, T.: Generalizations of Kempe's Universality Theorem. Master's thesis, Massachusetts Institute of Technology (2008)
2. Asimow, L., Roth, B.: The rigidity of graphs. Trans. Am. Math. Soc. **245**, 279–289 (1978)
3. Cruickshank, J., Guler, H., Jackson, B., Nixon, A.: Rigidity of linearly constrained frameworks. Int. Math. Res. Not. **12**, 3824–3840 (2020)
4. Eftekhari, Y., Jackson, B., Nixon, A., Schulze, B., Tanigawa, S., Whiteley, W.: Point-hyperplane frameworks, slider joints, and rigidity preserving transformations. J. Comb. Theory Ser. B **135**, 44–74 (2019)
5. Graver, J., Servatius, B., Servatius, H.: Combinatorial Rigidity. Graduate Studies in Mathematics, AMS, Providence, RI (1993)
6. Hewetson, J.: Recursive combinatorial constructions and rigidity of frameworks. Ph.D. Thesis, Lancaster University (2023)
7. Laman, G.: On graphs and rigidity of plane skeletal structures. J. Eng. Math. **4**, 331–340 (1970)
8. Lee, A., Streinu, I.: Pebble game algorithms and sparse graphs. Discrete Math. **308**(8), 1425–1437 (2008)
9. Nixon, A., Owen, J., Power, S.: Rigidity of frameworks supported on surfaces. SIAM J. Discret. Math. **26**(4), 1733–1757 (2012)
10. Nixon, A., Owen, J., Power, S.: A characterization of generically rigid frameworks on surfaces of revolution. SIAM J. Discret. Math. **28**(4), 2008–2028 (2014)
11. Pogorelov, A.V.: Extrinsic geometry of convex surfaces. Translation of the 1969 edition, Translations of Mathematical Monographs, vol. 35. AMS (1973)
12. Pollaczek-Geiringer, H.: Uber die Gliederung ebener Fachwerke. ZAMM-J. Appl. Math. Mech./Zeitschrift fur Angewandte Mathematik und Mechanik, **7** (1927), 58–72 and **12** (1932), 369–376
13. Saliola, F., Whiteley, W.: Some notes on the equivalence of first-order rigidity in various geometries (2007). arXiv:0709.3354
14. Tay, T.-S., Whiteley, W.: Generating isostatic frameworks. Struct. Topol. **11**, 21–69 (1985)

Managing Time Expanded Networks: The Strong Lift Problem

José-L. Figueroa, Alain Quilliot, Hélène Toussaint, and Annegret Wagler

Abstract Time Expanded Networks, built by considering the vertices of a base network all over some time space, are powerful tools for the formulation of problems that simultaneously involve resource assignment and scheduling. Still, in most cases, deriving algorithms from those formulations is difficult. We implement here a generic *Project and Lift* decomposition scheme while solving the *Strong Lift* issue, which consists in turning a solution defined on the base network into a solution of the whole problem with identical cost.

1 Introduction

A *Time Expanded Network* (TEN) $N^{\mathbf{TIME}}$ (see [6]) is derived from a network $N = (X, A)$ and a time space **TIME** according to the following construction: vertices of $N^{\mathbf{TIME}}$ are the copies (x, t) of the vertices x of N at the different instants t of **TIME**. An arc of $N^{\mathbf{TIME}}$ is either an *active* arc $\big((x, t), (y, t + \delta(t))\big)$, where $\delta(t)$ is the time required to traverse the arc (x, y) of N while starting at time t, or a *waiting* arc $\big((x, t), (x, t')\big)$ with $t < t'$, which expresses some kind of standby in x from time t to time t'. Note that **TIME** may be discrete or continuous.

TENs are powerful tools for modeling problems which simultaneously involve routing, scheduling, and synchronization mechanisms. The notion was first intro-

J.-L. Figueroa (✉) · A. Quilliot · H. Toussaint · A. Wagler
Université Clermont Auvergne, CNRS, Clermont Auvergne INP, Mines Saint-Etienne, LIMOS, 63000 Clermont-Ferrand, France
e-mail: jfg77_sigma@hotmail.com

A. Quilliot
e-mail: alain.quilliot@uca.fr

H. Toussaint
e-mail: helene.toussaint@uca.fr

A. Wagler
e-mail: annegret.wagler@uca.fr

A. Brieden et al. (eds.), *Graphs and Combinatorial Optimization: from Theory to Applications*, AIRO Springer Series 13,
https://doi.org/10.1007/978-3-031-46826-1_5

duced by Ford and Fulkerson [6] in order to show how problems involving both routing and scheduling could be cast into the network flow framework. It next gave rise to *dynamic networks* and *flow over time* models, where flow values may be trajectories. A trajectory means here a function from the time space **TIME** onto real or integer numbers, subject to constraints like continuity or Lipchitz inequalities. In the years 1990/2000, there were important contributions about algorithm design and applications of these notions to evacuation planning (see [1, 5, 7]). Insights were brought about the link between the TEN framework and the network flow over time models. More recently, some authors tried to combine the TEN framework with the improvement of the *Mixed Integer Linear Programming* (MILP) libraries in order to directly address some transportation problems (see [2, 3, 8]). Finally, we recently described the *Project/Lift* decomposition scheme at the core of the present contribution, and addressed the *Project* issue through Branch-and-Cut (see [4]).

The TEN construction does not provide us with natural algorithmic solutions. The fact is not only that the size of a TEN $N^{\mathbf{TIME}}$ increases very fast with the size of the time space **TIME**, but also that trying to control this size through rounding tends to induce uncontrolled error propagation. So, our purpose here is to bypass those difficulties by implementing the following *Project /Lift* decomposition scheme: We first solve (*Project* step) a projection of our problem set onto the base graph N, and next try (*Lift* step) to turn this projected solution into a full feasible solution defined on the TEN $N^{\mathbf{TIME}}$. Since we formerly addressed the *Project* issue (see [4]) we focus here on the *Lift* issue. We address it in the "strong" way that means while imposing that the projection of the lifted solution is exactly equal to the projected solution.

In order to make our methods easier to understand and to perform numerical experiments, we refer inside this paper to a model related to the management of an *item relocation* process involving a set of carriers.

The paper is organized as follows. In Sect. 2, we present a TEN model for our reference problem together with the projected model which derives from this TEN model. In Sects. 3–4, we address the *Strong Lift* issue and perform some numerical experiments.

2 A Reference TEN Relocation Model

We consider here a *transit network* $N = (X, A)$, together with a distinguished vertex $Depot$. Every arc $a = (x, y)$ of N is provided with a *time value* $T_{(x,y)}$ and with a *cost value* $C_{(x,y)}$. We use the following notations:

- We set $\mathbf{T} = (T_{(x,y)} : (x, y) \in A)$ and $\mathbf{C} = (C_{(x,y)} : (x, y) \in A)$. For any path π from $x \in X$ to $y \in X$, we denote by $L^T(\pi)$ its length in the sense of $\mathbf{T}$. We do the same with $\mathbf{C}$. For any pair of vertices (x, y) we denote by $D^T(x, y)$ the shortest path distance induced by $\mathbf{T}$ from x to y, and by $D^C(x, y)$ the shortest path distance induced by $\mathbf{C}$ from x to y.

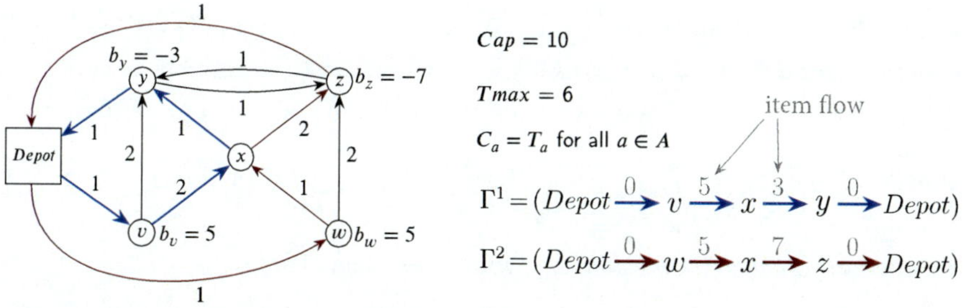

Fig. 1 The transit network $N = (X, A)$ used in *Example 1*

- Let U be some subset of X. We set $\partial_N^-(U) = \{(x, y) \in A : x \notin U,\ y \in U\}$, $\partial_N^+(U) = \{(x, y) \in A : x \in U,\ y \notin U\}$, $\partial_N(U) = \partial_N^-(U) \cup \partial_N^+(U)$, and $A(U) = \{(x, y) \in A : x \in U,\ y \in U\}$. We simplify these notations in case of a singleton $\{x\}$ by writing $\partial_N^-(x)$, $\partial_N^+(x)$ and $\partial_N(x)$ instead of $\partial_N^-(\{x\})$, $\partial_N^+(\{x\})$ and $\partial_N(\{x\})$, respectively.

Items are located inside the network and must be relocated, within a discrete time horizon $\{0, 1, \dots, T_{\max}\}$ by a fleet of identical carriers with capacity Cap. We are provided with an integral *balance* vector $\mathbf{b} = (b_x, x \in X)$ such that $\sum_{x\in X} b_x = 0$. A value $b_x > 0$ means that x is an *excess vertex* and that b_x items must be removed from x; a value $b_x < 0$ means that x is a *deficit vertex* and that $-b_x$ items must be brought to x. The *Item Relocation Problem* (IRP) consists in scheduling the carriers in such a way they perform this relocation process while meeting the time horizon and carrier capacity requirements and while minimizing a hybrid cost $\alpha \cdot c_1 + \beta \cdot c_2 + \gamma \cdot c_3$, where c_1 is the number of active carriers, c_2 is their running cost in the sense of $\mathbf{C}$, c_3 is the time that items spend moving inside the carriers, and α, β, γ are scaling coefficients. We allow *preemption*, which means that carriers may exchange items during the process.

Example 1 The network $N = (X, A)$ depicted in Fig. 1 shows two carrier routes Γ^1 and Γ^2. Note that two items are transferred from the carrier in Γ^1 to the carrier in Γ^2 at the vertex x, and we have that $c_1 = 2$, $c_2 = 10$, and $c_3 = 32$. □

2.1 A TEN Relocation Commodity Flow Model

In order to cast the IRP into the TEN framework (see [8]), we first derive from the transit network $N = (X, A)$ its time expansion $N^{T_{\max}} = (X^{T_{\max}}, A^{T_{\max}})$ according to $T_{\max}$. The vertex set $X^{T_{\max}}$ is the set of all pairs (x, t), $x \in X$, $t \in \{0, 1, \dots, T_{\max}\}$, augmented with two distinguished vertices *source* and *sink*. The arcs $a \in A^{T_{\max}}$ are classified as follows, together with their *carrier cost* $\hat{C}_a$, and their *item cost* $\hat{I}_a$:

- *input-arcs* $a = \big(source, (x, 0)\big)$, $x \in X$, with $\hat{I}_a = 0$ and $\hat{C}_a = 0$;
- *output-arcs* $a = \big((x, T_{\max}), sink\big)$, $x \in X$, with $\hat{I}_a = \hat{C}_a = 0$;
- *waiting-arcs* $a = \big((x, t), (x, t + 1)\big)$, $x \in X$, $t \in \{0, \ldots, T_{\max} - 1\}$, with $\hat{I}_a = \hat{C}_a = 0$;
- *active-arcs* $a = \big((x, t), (y, t + T_{(x,y)})\big)$, $(x, y) \in A$, $t \in \{0, \ldots, T_{\max} - T_{(x,y)}\}$, with $\hat{I}_a = \gamma \cdot T_{(x,y)}$ and $\hat{C}_a = \beta \cdot C_{(x,y)}$;
- *backward-arc* $a = (sink, source)$, with $\hat{I}_a = 0$ and $\hat{C}_a = \alpha$.

Now we formalize the IRP as a 2-commodity flow model on $N^{T_{\max}}$.

TEN IRP Formulation. Compute two nonnegative integral $A^{T_{\max}}$-indexed vectors **H** and **h** (for carriers and items, respectively) such that:

- **H** and **h** satisfy flow conservation at any vertex of $X^{T_{\max}}$; (E1)
- for any active-arc $a = \big((x, t), (y, t + T_{(x,y)})\big)$: $h_a \leq Cap \cdot H_a$; (E2)
- for any input-arc $a = \big(source, (x, 0)\big)$, $x \neq Depot$:
  $H_a = 0$; $h_a = \max(b_x, 0)$; (E3)
- for any output-arc $a = \big((y, T_{\max}), sink\big)$, $y \neq Depot$:
  $H_a = 0$; $h_a = \max(-b_y, 0)$; (E4)
- the global cost $Cost(H, h) = \sum_{a \in A^{T_{\max}}} \big(H_a \cdot \hat{C}_a + h_a \cdot \hat{I}_a\big)$ is minimized.

(E1) expresses the circulation of carriers and items inside N. (E2) ensures that any item moving between two vertices x and y is contained into some carrier. Constraints (E3) and (E4) characterize initial and final states.

Figure 2 shows the construction of the TEN $N^{T_{\max}} = (X^{T_{\max}}, A^{T_{\max}})$ associated to the network $N = (X, A)$ of Fig. 1 and $T_{\max} = 6$. It turns the solution of *Example 1* into a 2-commodity flow vector (**H**, **h**).

2.2 The Projected *IRP* Model

A flow vector **H** being given on $N^{T_{\max}} = (X^{T_{\max}}, A^{T_{\max}})$, we define its *projection* **F** on N by setting, for any arc (x, y) of N: $F_{(x,y)} = \sum_{t=0}^{T_{\max}} H_{(x,t),(y,t+T_{(x,y)})}$. We define the *projection* **f** of **h** exactly the same way, and we call (**F**, **f**) the *projection* of (**H**, **h**) on the network N.

F and **f** must be such that:

- **F** satisfies flow conservation at any vertex of X; (E5.1)
- for any vertex x of N: $\sum_{a \in \partial_N^+(x)} f_a - \sum_{a \in \partial_N^-(x)} f_a = b_x$; (E5.2)
- for any arc a of N: $f_a \leq Cap \cdot F_a$; (E6)

Those constraints are not enough to characterize **F** and **f**: They do not forbid subtours and they do not provide us with a well-fitted

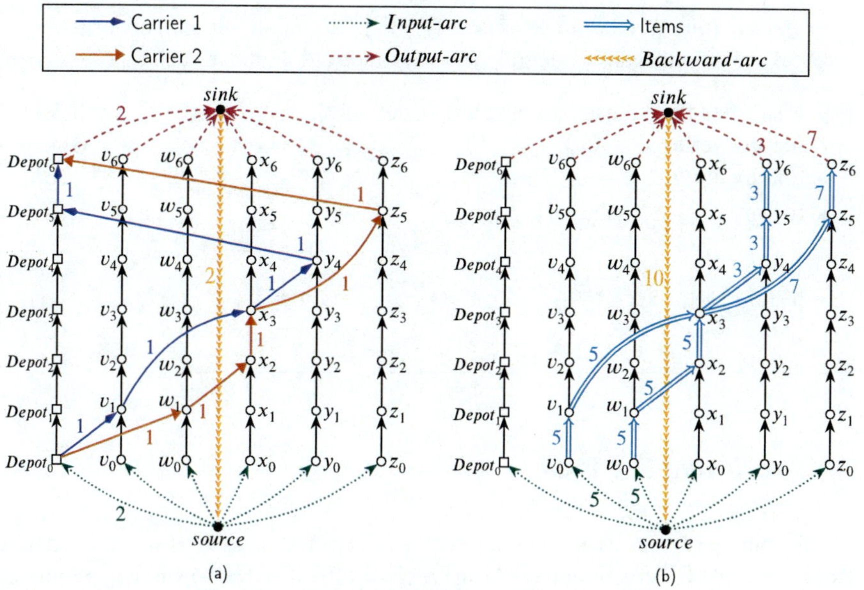

Fig. 2 The routes and schedules from *Example 1* viewed as a 2-commodity flow on the TEN $N^{T_{\max}} = (X^{T_{\max}}, A^{T_{\max}})$. **a** Carrier flow vector **H**. **b** Item flow vector **h**

estimation of the carrier number c_1. In order to enhance our projected model, we first notice as in [4] that the quantity $\sum_{a \in A} T_a \cdot F_a$ provides us with the global time that carriers spend running inside N. Since the whole process must be performed in no more than $T_{\max}$ time units, we need at least $\lceil (\sum_{a \in A} T_a \cdot F_a)/T_{\max} \rceil$ carriers. As a consequence, $(\mathbf{F}, \mathbf{f})$ should minimize the *projected cost*: (E7)

$$PCost(\mathbf{F}, \mathbf{f}) = \alpha \cdot \frac{\left(\sum_{a \in A} T_a \cdot F_a\right)}{T_{\max}} + \beta \cdot \left(\sum_{a \in A} C_a \cdot F_a\right) + \gamma \cdot \left(\sum_{a \in A} T_a \cdot f_a\right).$$

Similarly we notice that for any $U \subset X \setminus \{Depot\}$, the time carriers spend at the border or inside U is equal to $\sum_{a \in \partial_N(U) \cup A(U)} T_a \cdot F_a$. For any carrier q this time must not exceed $T_{\max}$. Since the number of carriers that serve the vertices of U is $\sum_{a \in \partial_N^-(U)} F_a$, we deduce that $\mathbf{F}$ must satisfy the following *Extended Subtour* constraint:

$$T_{\max} \cdot \left(\sum_{a \in \partial_N^-(U)} F_a\right) \geq \sum_{a \in \partial_N(U) \cup A(U)} T_a \cdot F_a. \quad \text{(E8)}$$

One may check that (E8) can be separated in polynomial time (see [4]). This allows us to set the following projected model which can be handled by Branch-and-Cut:

Projected Item Relocation Problem (PIRP). Compute on the network $N = (X, A)$ two nonnegative integral vectors A-indexed **F** and **f** such that:

- **F** satisfies flow conservation at any vertex of X; (E5.1)
- for any vertex $x \in X$, $\sum_{a \in \partial_N^-(x)} f_a - \sum_{a \in \partial_N^+(x)} f_a = b_x$; (E5.2)
- for any arc $a \in A$, $f_a \leq Cap \cdot F_a$; (E6)
- for any $U \subseteq X \setminus \{Depot\}$, $T_{\max} \cdot \sum_{a \in \partial_N^-(U)} F_a \geq \sum_{a \in \partial_N(U) \cup A(U)} T_a \cdot F_a$; (E8)
- Minimize $\alpha \cdot \frac{\left(\sum_{a \in A} T_a \cdot F_a\right)}{T_{\max}} + \beta \cdot \left(\sum_{a \in A} C_a \cdot F_a\right) + \gamma \cdot \left(\sum_{a \in A} T_a \cdot f_a\right)$. (E7)

3 The *Strong Lift* Issue

The former section raises in a natural way the following *Lift* issue: How can we derive from a feasible PIRP solution (**F**, **f**) an efficient TEN IRP (**H**, **h**) while applying the following decomposition scheme?

***Project/Lift* Decomposition Scheme.**

1. Solve the PIRP model and get a projected solution (**F**, **f**).
2. Turn (i.e., *lift*) (**F**, **f**) into a "good" solution (**H**, **h**) of the TEN IRP model.

3.1 The Strong Lift Model

The most natural way to formalize this *Lift* issue consists in asking (**H**, **h**) to be such that its projection onto N is exactly (**F**, **f**). This leads us to set the following *Strong Lift Problem*:

Strong Lift Problem SLIFT(**F**, **f**). Compute a feasible IRP solution (**H**, **h**) in such a way that:

- the projection of **H** (respectively, **h**) on the transit network N is equal to **F** (respectively, **f**);
- the cost value $Cost$(**H**, **h**) is smallest possible.

If (**H**, **h**) is a feasible solution of SLIFT(**F**, **f**) then the difference between $Cost$(**H**, **h**) and $PCost$(**F**, **f**) only reflects the difference between the true number of carriers $\mathbf{H}_{(sink,source)}$ and its approximation $(\sum_{a \in A} T_a \cdot F_a)/T_{\max}$ as expressed in the PIRP Model.

3.2 A Necessary Condition for the Feasibility of the Strong Lift Problem: Enhancing the PIRP Model

A PIRP solution (**F**, **f**) may not always be *liftable*. So we should try to reinforce this projected model in order to enhance the probability that (**F**, **f**) becomes liftable. We intend to do it in such a way that (**F**, **f**) becomes "partially" liftable, that means that there exists a feasible TEN IRP solution (**H**, **h**) such that the projection of **h** is **f**. In order to set this in a formal way, we need to introduce the notion of a *feasible path*.

- A *feasible path* (seen as a set of arcs) of N is any path π from an excess vertex x to a deficit vertex y whose length $L^T(\pi)$ in the sense of the time matrix **T** satisfies: $D^T(Depot, x) + L^T(\pi) + D^T(y, Depot) \leq T_{\max}$. We associate, with any such feasible path, a flow vector $\mathbf{f}^\pi$ which transports one item from x to y along the path π. We denote by Π^{FP} the set of feasible paths.
- A flow vector **f** is *feasible-path-decomposable* if and only if it can be written as $\mathbf{f} = \sum_{\pi \in \Pi^{FP}} \lambda_\pi \, \mathbf{f}^\pi$, with $\lambda_\pi \in \mathbb{R}_+$ for all $\pi \in \Pi^{FP}$.
- We say that an A-indexed vector **w** is a *Path Feasibility* vector if, for any feasible path π, we have $\sum_{a \in \pi} w_a \geq 0$.

An item starting from an excess vertex x can be transported to some deficit vertex y along path π only if π is a feasible path. It follows that a solution (**F**, **f**) of the PIRP Model may be lifted into a feasible IRP solution (**H**, **h**) only if **f** is feasible-path-decomposable. This leads us to the following necessary condition for the feasibility of the *Strong Lift Problem*.

Theorem 1 *The Strong Lift Problem SLIFT(F, f) admits a feasible solution if and only if, for any Path Feasibility vector* **w**, *the following inequality holds:*
$\sum_{a \in A} f_a \cdot w_a \geq 0$. *(E10)*

Proof Necessity is straightforward from the above explanation. As for the sufficiency, we get it by noticing that (E10) is nothing more than a formulation of Farkas lemma in the case of vector **f** and the vector collection $\{\mathbf{f}^\pi, \pi \in \Pi^{FP}\}$: A flow vector **f** is feasible-path-decomposable if and only if it belongs to the cone defined by the collection $\{\mathbf{f}^\pi, \pi \in \Pi^{FP}\}$, that means (Farkas lemma) if and only if for any vector **w** whose scalar product $\sum_{a \in A} \mathbf{f}^\pi_a \cdot w_a \geq 0$ with any vector $\mathbf{f}^\pi$ is nonnegative, then the scalar product $\sum_{a \in A} \mathbf{f}_a \cdot w_a$ is also nonnegative.

The latter result suggests us to enhance our projected model with the following *Feasible Path* constraint:

- For any *Path Feasibility* vector **w**: $\sum_{a \in A} f_a \cdot w_a \geq 0$. (E10)

One easily checks that separating (E10) may be performed in practice through a simple column generation process. It follows that the resulting augmented PIRP may be handled through Branch-and-Cut.

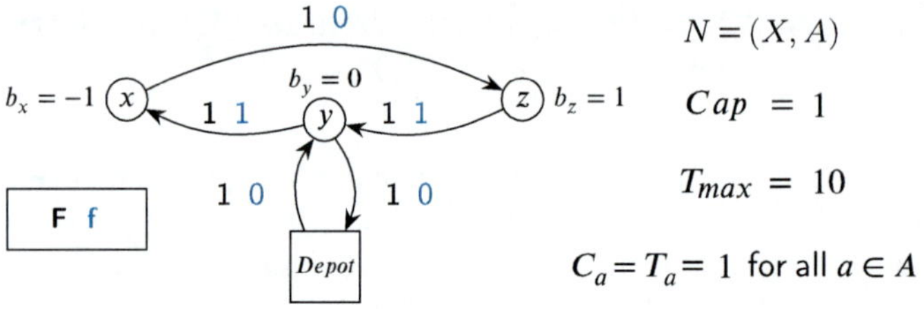

Fig. 3 A PIRP solution (**F**, **f**) on $N = (X, A)$ that satisfies (E10) but cannot be lifted

Remark 1 (E10) is not a sufficient condition for (F, f) to be liftable. Figure 3 shows that even if we impose the *Feasible Path* constraints, a PIRP solution (**F**, **f**) cannot always be viewed as the projection of a feasible solution (**H**, **h**) of IRP TEN. The carrier follows the route $(Depot, y, x, z, y, Depot)$, but cannot transport any item from z to x. In fact, it is known that computing (**H**, **h**) from (**F**, **f**) in such a way that (**F**, **f**) is the projection of (**H**, **h**) with identical cost value, is NP-Hard [3].

4 A MILP Formulation of the Strong Lift Problem

Let us recall that our strong version of the *Lift Problem* is about the search of an IRP solution (**H**, **h**) whose *projection* on the network N is exactly the solution (**F**, **f**) obtained through resolution of the projected PIRP model. So let us consider a feasible (optimal) solution (**F**, **f**) of the PIRP model. We denote by Q(**F**) the sum $\sum_{x \in X} \left((\sum_{a \in \partial_N^-(x)} F_a) \cdot (\sum_{a \in \partial_N^+(x)} F_a) \right)$, and by S(**F**) the sum $\sum_{a \in A} F_a$. We are going to show that it is possible to set an exact MILP formulation of the *Strong Lift Problem*, which involves $2 \cdot Q$(**F**) decision variables, together with Q(**F**) $+3 \cdot S$(**F**) rational load and time variables.

The idea is that solving the *Strong Lift Problem* basically means determining what happens *"inside"* the vertices of the transit network N: More precisely, a vertex x being given, we want to know along which arc a' a given carrier (respectively, item) which arrives into x along some arc a is going to leave x, and at which time.

4.1 Solving the Strong Lift Problem in an Exact Way

In order to formalize this idea, we construct a network $Strong(N, \mathbf{F})$.

- With any arc $a = (x, y) \in A$ such that $F_a \geq 1$, we associate the set $Copy(a)$ of F_a *copy-arcs* $a^m, m = 1, \ldots, F_a$, with respective origin vertices $p = (x, a, m, +)$, and respective destination vertices $q = (y, a, m, -)$. It follows that, at the same time we create those copy-arcs, we also create *copy-vertices* $p = (x, a, m, +)$ and

$q = (y, a, m, -)$, which respectively correspond to the carriers which leave x and to the carriers which arrive into y. We denote by X^* the resulting vertex set and by $Copy(A)$ the set of all copy-arcs. For any such vertex $p = (y, a, m, \varepsilon)$, we set $x(p) = y$ and $\varepsilon(p) = \varepsilon$, and, for any vertex y of N, we set:

- $X^*(y) = \{p \in X^* \text{ such that } x(p) = y\}$;
- $X^*Plus(y) = \{p \in X^* \text{ such that } x(p) = y,\ \varepsilon(p) = +\}$;
- $X^*Minus(y) = \{p \in X^* \text{ such that } x(p) = y,\ \varepsilon(p) = -\}$;
- $CopyIn(y) = \{a \in Copy(A) \text{ with destination in } X^*Minus(y)\}$;
- $CopyOut(y) = \{a \in Copy(A) \text{ with origin in } X^*Plus(y)\}$.

- We complete the arc collection $\{a^m, a = (x, y) \text{ such that } F_a \geq 1, m = 1, \ldots, F_a\}$ by *middle-arcs* $u = ((x, a, m, -), (x, a', m', +))$ which, for any vertex x of N, connect any copy-vertex $(x, a, m, -)$, where a has destination x, $m = 1, \ldots, F_a$, to any copy-vertex $(x, a', m', +)$, where a' has origin x, $m' = 1, \ldots F_{a'}$. We denote by $Middle$ the set of all middle-arcs created that way, and, for any x, we denote by $Middle(x)$ the set of the middle-arcs u whose origin may be written $(x, a, m, -)$. Notice that $Middle(x)$ defines a complete bipartite graph on $X^*(x)$. For any vertex $p = (x, a, m, +)$, we denote by $In(p)$ the set of middle-arcs u whose destination is p, and, for any vertex $p = (x, a, m, -)$, we denote by $Out(p)$ the set of middle-arcs u whose origin is p. For any vertex x of N, we set $MiddleOut(x) = \bigcup_{p \in X^*Minus(x)} Out(p)$, and $MiddleIn(x) = \bigcup_{p \in X^*Plus(x)} In(p)$.

We denote by $Strong(N, \mathbf{F})$ the resulting network (see Fig. 4), which contains $2 \cdot S(\mathbf{F})$ vertices, $S(\mathbf{F})$ copy-arcs, and $Q(\mathbf{F})$ middle-arcs.

To set up our SLIFT($\mathbf{F}$, $\mathbf{f}$) model, we use the following variables:

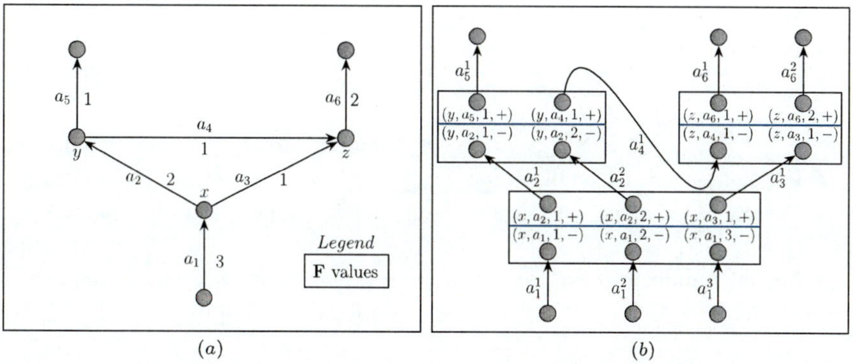

Fig. 4 Constructing the graph $Strong(N, \mathbf{F})$. **a** A set of arcs in a network N together with their corresponding $\mathbf{F}$ flow values. **b** The arcs and vertices in the graph $Strong(N, \mathbf{F})$ that are created from the arcs, vertices, and flow values in (**a**). To avoid a cumbersome drawing we have not depicted the arcs in the set $Middle$

- $\mathbf{Z} = (Z_u, u = ((x, a, m, -), (x, a', m', +)) \in Middle)$, with $\{0, 1\}$ values, where $Z_u = 1$ means that the carrier which arrives at vertex x along copy-arc a^m keeps on along arc $a'^{m'}$.
- $\mathbf{z} = (z_u, u = ((x, a, m, -), (x, a', m', +)) \in Middle)$, with rational values: z_u may be kept integral and so may be viewed as the number of items which arrive at vertex x along arc a^m and which are transferred to arc $a'^{m'}$.
- $\ell = (\ell_u, u = ((x, a, m, -), (x, a', m', +)) \in Middle)$, with $\{0, 1\}$ values: where $\ell_u = 1$ means that $z_u \neq 0$.
- $\mathbf{z}^* = (z^*_{a^m}, a^m \in Copy(A))$ with rational values: similarly as for z, z_{a^m} shall correspond to the number of items transported along arc a^m;
- $\mathbf{t} = (t_p, p = (x, a, m, \varepsilon) \in X^*)$ with rational nonnegative values: t_p stands for the time when a carrier arrives (in case $\varepsilon = -$) or leaves (in case $\varepsilon = +$) in x along arc a^m.

MILP model SLIFT(F, f). Compute on the network $Strong(N, \mathbf{F})$ two 0–1 vectors $Middle$-indexed $\mathbf{Z}$ and ℓ; one nonnegative integral vector $Middle$-indexed $\mathbf{z}$; one nonnegative integral vector $Copy(A)$-indexed $\mathbf{z}^*$; and one rational nonnegative vector X^*-indexed $\mathbf{t}$, such that:

- For any copy-vertex $q = (x, a, m, \varepsilon)$ of $Strong(N, \mathbf{F})$, with $x \neq Depot$: $\sum_{u \in MiddleIn(q)} Z_u = 1 = \sum_{u \in MiddleOut(q)} Z_u$. (E11.1)
- For any copy-vertex $q = (Depot, a, m, +)$: $\sum_{u \in MiddleIn(q)} Z_u \leq 1$. (E11.2)
- For any copy-vertex $p = (Depot, a, m, -)$: $\sum_{u \in MiddleOut(p)} Z_u \leq 1$. (E11.3)
- For any middle-arc u: $z_u \leq Cap \cdot \ell_u$. (E12.1)
- For any copy-arc a^m: $z^*_{a^m} \leq Cap$. (E12.2)
- For any vertex $q = (y, a, m, -)$: $z^*_{a^m} \geq \sum_{u \in MiddleOut(q)} z_u$. (E13.1)
- For any vertex $p = (x, a, m, +)$: $z^*_{a^m} \geq \sum_{u \in MiddleIn(p)} z_u$. (E13.2)
- For any vertex x of N: $\sum_{u \in CopyIn(x)} z^*_u = \sum_{u \in MiddleOut(x)} z_u + \max(-b_x, 0)$. (E14.1)
- For any vertex x of N: $\sum_{u \in CopyOut(x)} z^*_u = \sum_{u \in MiddleIn(x)} z_u + \max(b_x, 0)$. (E14.2)
- For any arc $a = (x, y)$ of N: $\sum_{u \in Copy(a)} z^*_u = f_a$. (E15)
- For any *copy-arc* $(p, q) = \big((x, a = (x, y), m, +), (y, a = (x, y), m, -)\big)$: $t_q \geq t_p + T_{(x,y)}$. (E16)
- For any middle-arc $u = (q = (x, a, m, -), p = (x, a', m', +))$, the implication $\big((Z_u = 1) \vee (\ell_u = 1)\big) \Rightarrow t_p \geq t_q$ holds, equivalent to: $Z_u + \frac{t_q - t_p}{T_{\max}} \leq 1$ and $\ell_u + \frac{t_q - t_p}{T_{\max}} \leq 1$. (E17)

Objective function Maximize $\sum_{u \in Middle(Depot)} Z_u$. (E18)

The meaning of those constraints becomes clear from the proof of Theorem 2.

Theorem 2 *Solving the above MILP model* SLIFT(F, f)*, which involves* $2 \cdot Q(F)$ *decision variables* $\mathbf{Z}$ *and* ℓ*, together with* $Q(F) + 3 \cdot S(F)$ *rational variables* z*,* z^* *and* $\mathbf{t}$*, also solves the Strong Lift Problem related to* (F, f) *in an exact way.*

Proof This result derives from the fact that we require the projection of $\mathbf{H}$ onto network N to be exactly equal to $\mathbf{F}$. More precisely, since the *Strong Lift Problem* explicitly requires the projection of $\mathbf{H}$ onto the transit network N to be equal to $\mathbf{F}$, we see that the routes followed by the carriers are completely determined by the way we assign a carrier entering into a vertex x along some copy-arc a^m onto another copy-arc $a'^{m'}$ leaving x (in case $x = Depot$, we may assign a *"null" arc*, that means consider that the carrier ends its trip into $Depot$ with the arc a^m). Decision vector $\mathbf{Z}$, together with matching constraints (E11.1–E11.3) express the way carrier routes distribute themselves inside any vertex x. As for the items, we first notice that once $\mathbf{Z}$ has been computed, $\mathbf{z}$ and $\mathbf{z}^*$ come as the solution of a Min-Cost Flow problem. So SLIFT$(\mathbf{F}, \mathbf{f})$ behaves as if both vectors were imposed to be integral. Since any item move from x to y must be covered by some carrier, any item arriving to some vertex x_i along some copy-arc a^m will have either to remain in x as part of the negative deficit b_x or keep on along another copy-arc $a'^{m'}$ leaving x. Constraints (E13.1 - E15) express the way items are going to distribute themselves while traversing this vertex x. Deriving an IRP solution $(\mathbf{H}, \mathbf{h})$ from a PIRP solution $(\mathbf{F}, \mathbf{f})$ and from vectors $\mathbf{Z}$, ℓ, $\mathbf{z}$, $\mathbf{z}^*$, becomes possible if we are able to embed the vertices of the $Strong(\mathbf{F}, \mathbf{f})$ graph into the Time Expanded Network $N^{T_{\max}}$, that means if we can compute a vector $\mathbf{t}$ which meets constraints (E16, E17). It follows that any feasible solution of the *Strong Lift Problem* may be turned into a feasible solution of SLIFT$(\mathbf{F}, \mathbf{f})$ and conversely. We conclude by noticing that the value of the objective function $\sum_{u \in Middle(Depot)} Z_u$ is merely the difference between the value $\sum_x F_{(Depot,x)}$ and the number of carriers, while the other components of the cost IRP function are the same for $(\mathbf{H}, \mathbf{h})$ and $(\mathbf{F}, \mathbf{f})$. It follows that solving SLIFT$(\mathbf{F}, \mathbf{f})$ makes us minimize the number of carriers involved into the lifted solution $(\mathbf{H}, \mathbf{h})$ whose projection onto N is exactly $(\mathbf{F}, \mathbf{f})$.

4.2 Numerical Experiments

We performed several numerical experiments, whose purpose is to estimate the feasibility of the *Strong Lift Model* and the gap between the number of vehicles obtained after resolution of this model and its approximation according to the PIRP model. We ran those experiments on a computer with a 2.3 GHz Intel Core i5 processor and 16 GB RAM, while using the C++ language (compiled with *Apple Clang 10*) and the CPLEX12.10 MILP library.

No standardized benchmarks exist for the generic IRP. So we built instances as follows: The station set X is a set of n points inside a 100×100 grid, the set of arcs A consists of m arcs generated randomly, the time matrix $\mathbf{T} = (T_{(x,y)}, (x, y) \in A)$ corresponds to the rounded Euclidean Distance and the cost matrix

Table 1 Strong lift numerical results

Id	n	m	Cap	T_{max}	λ	α	β	γ	PIRP	VPIRP	TPIRP	SL	VSL	TSL
1	20	78	2	324	4	304	1.0	1.000	2110.85	3	1.61	–	–	0.01
2	20	65	5	400	5	150	0.4	0.500	1196.10	3	0.01	–	–	0.01
3	20	50	5	603	9	392	0.4	0.250	2354.43	3	0.44	2474.70	3	0.01
4	20	62	5	420	6	300	0.4	0.500	2727.30	4	0.32	–	–	0.01
5	50	155	5	390	6	196	0.4	0.500	4326.10	7	0.01	–	–	0.01
6	50	146	20	436	4	312	0.1	0.125	1840.17	4	19.33	–	–	0.01
7	50	217	5	912	8	672	0.4	0.250	1643.76	2	124.47	2153.65	2	0.14
8	100	363	2	336	4	252	1.0	1.000	17179.00	22	81.58	–	–	0.01
9	100	289	10	432	4	360	0.2	0.250	3272.98	4	70.22	–	–	0.01
10	100	327	5	552	8	392	0.5	0.200	5944.23	7	46.90	–	–	1630.87

C= $(C_a, a \in A)$ to the Manhattan Distance. Each vertex x but *Depot* is assigned a b_x value in $\{-10, \ldots, 10\}$, the capacity *Cap* belongs to $\{2, 5, 10, 20\}$, the time horizon limit $T_{\max}$ is the product $\lambda \cdot (\max_{(x,y)\in A} T_{(x,y)})$ with $\lambda \in \{4, 5, 6, 8, 9\}$. The scaling coefficients α, β, γ are chosen in such a way that the values of cost components $\alpha \cdot$ *number of carriers*, $\beta \cdot$ *carrier ride cost* and $\gamma \cdot$ *items ride time* become comparable. The first nine columns of Table 1 summarize those characteristics.

The same Table 1 displays the output values of the SLIFT(**F**, **f**) MILP. Column **PIRP** corresponds to the optimal value (with respect to the objective function (E7)) of the projected PIRP model, **VPIRP** shows the estimated number of carriers (related to **PIRP**), and **TPIRP** the related running time (in seconds). The column **SL** displays the optimal value of SLIFT(**F**, **f**), **VSL** the related number of carriers, and **TSL** indicates the related CPU time (in seconds). Missing values are indicated by a hyphen symbol–, and correspond to PIRP solutions (**F**, **f**) for which the corresponding SLIFT(**F**, **f**) MILP is infeasible.

We see that solving SLIFT(**F**, **f**) can be done in a reasonable computing time. But we also see that in many cases this model happens to be infeasible. This means that we should accept, while dealing with the *Lift* issue, a deterioration of the cost induced by the projected model. We shall address this requirement in a future work.

References

1. Aronson, J.: A survey of dynamic network flows. Ann. Oper. Res. **20**, 1–66 (1989)
2. Bsaybes, S., Quilliot, A., Wagler, A.: Fleet management for autonomous vehicles using flows in time-expanded networks. TOP **27**(2), 288–311 (2019) (Springer Verlag)
3. Chemla, D., Meunier, F., Wolfler-Calvo, R.: Bike sharing systems: solving the static rebalancing problem. Disc. Opt. **10**(2), 120–146 (2013)
4. Figueroa González, J.L., Baïou, M., Quilliot, A., Toussaint, H., Wagler, A.: Branch-and-cut for a 2-commodity flow relocation model with time constraints. ISCO 2022. LNCS, vol. 13526. Springer, Berlin (2022)
5. Fleischer, L., Skutella, M.: Quickest flow over time. SIAM J. Comput. **36**(6), 1600–1630 (2007)
6. Ford, R.L., Fulkerson, D.R.: Flows in networks. Princeton, NJ (1962)
7. Hall, A., Hippler, S., Skutella, M.: Multicommodity flows over time. Theor. Comput. Sci. 58–84 (2007)
8. Krumke, S., Quilliot A., Wagler A., Wegener, J.: Relocation in carsharing systems using flows in time-expanded networks. In: Gudmundsson, J., Katajainen, J. (eds.) Experimental Algorithms, pp. 87–98. Springer, Berlin (2014)
9. Raviv, T., Tzur, M., Forma, I.A.: Static repositioning in a bike-sharing system. EURO J. Transp. Logist. **2**, 187–229 (2013)

k-Slow Burning: Complexity and Upper Bounds

Michaela Hiller, Arie M. C. A. Koster, and Philipp Pabst

Abstract The graph burning problem studies the speed at which information can spread in graphs across their edges. We discuss a recently introduced variant of the problem, k-slow burning, in which every burning vertex can only ignite up to k of its neighbours in each step of the burning process. We consider the complexity of computing the corresponding graph parameter, the k-slow burning number $b_s(k, G)$. We prove $\mathcal{NP}$-hardness on multiple graph classes, most notably the class of graphs of radius 1, where normal graph burning is solvable in polynomial time. Furthermore, we show that among all connected graphs on n vertices, the burning number of the star graph, $b_s(k, S_{n-1})$, is maximal for $k \in \{1, 2\}$ and asymptotically maximal for fixed $k \geq 3$. This observation leads to a generalisation of the burning number conjecture in regard to k-slow burning.

1 Introduction

The notion of *graph burning* was introduced as a model for contagion in social networks [3] and has since been the subject of extensive research. Topics of interest include the computational complexity of the problem (e.g., [1, 12]), approximation algorithms (e.g., [5, 7, 9]) and upper bounds for the associated graph parameter, the burning number $b(G)$. An overview of results can be found in [2]. Graph burning is carried out as a step-wise process on an undirected graph $G = (V, E)$, $|V| = n$, where in every step first, every burning vertex spreads the fire to its entire neighbourhood, before second, a new *source of fire* is ignited. If $(v_1, \ldots, v_t)$ is a sequence, such

M. Hiller · A. M. C. A. Koster
Discrete Optimization, RWTH Aachen, Aachen, Germany
e-mail: hiller@math2.rwth-aachen.de

A. M. C. A. Koster
e-mail: koster@math2.rwth-aachen.de

P. Pabst (✉)
Chair for Management Science, RWTH Aachen, Aachen, Germany
e-mail: philipp.pabst@oms.rwth-aachen.de

A. Brieden et al. (eds.), *Graphs and Combinatorial Optimization: from Theory to Applications*, AIRO Springer Series 13,
https://doi.org/10.1007/978-3-031-46826-1_6

that, when choosing v_i as the ith source of fire, G can be burned within t time steps, it is called a *burning sequence* for G. The aim is to choose the new sources of fire in a way that minimises the length of a burning sequence. The minimum number of steps necessary to ignite every vertex in G is denoted by the *burning number*, $b(G)$.

The central open question within the area of graph burning is given by the *burning number conjecture* [3], which states that $b(G) \leq \lceil \sqrt{n} \rceil$ for all connected graphs G on n vertices. This conjectured upper bounds is reached for paths P_n and cycles C_n on n vertices. There are numerous results either proving upper bounds close to $\lceil \sqrt{n} \rceil$ (e.g., [6, 15]) or showing the conjecture to hold for certain classes of graphs, such as spiders or p-caterpillars, $p \leq 2$ [7, 12], however the conjecture itself remains open.

Graph burning is placed alongside a multitude of other problems studying the spread of information across networks. One of the oldest such problems is the *(minimum time) k-broadcasting* problem [11, 16], where, starting from a single vertex, some information is distributed across a graph. Here, in each time step every informed vertex spreads the information to up to k of its neighbours. However, in contrast to the second step in graph burning, no new *sources of information* are chosen in subsequent time steps. Thus, in order to minimise the number of time steps needed to inform the entire graph, an optimal *broadcast protocol* has to optimise the order in which every vertex informs its neighbours once it is informed itself. The problem is $\mathcal{NP}$-hard in general [8] and polynomially solvable on trees [10].

For certain applications neither the graph burning model nor k-broadcasting is realistic. E.g., consider some political party campaigning for voters. In every time step, the party can hire a new campaigner, represented as a source of fire in the graph burning model. However, it is unrealistic to assume that this campaigner can convince all of their acquaintances within a single time step. Instead, we assume in our generalised model that they can only influence up to k of their acquaintances per time step, as in k-broadcasting.

To formalise this behaviour, in [2, 14] it was suggested to study a variant of the graph burning problem, k-slow burning, that forms a midpoint between graph burning and k-broadcasting. Here, in every time step, first every burning vertex spreads the fire to up to k of its neighbours (as for k-broadcasting) before second a new source of fire is chosen (as for graph burning). We define the terms *k-slow burning sequence* and *k-slow burning number*, denoted by $b_s(k, G)$, analogously as for normal graph burning. This leads to the following decision problem.

k- SLOW BURNING
Input: A Graph G, an integer k and a time bound t.
Question: Does $b_s(k, G) \leq t$ hold?

In order to minimise the length of a *k-slow burning sequence*, we have to optimise the choice of neighbours in the first step as well as the choice of new sources of fire in the second step of the process. This observation suggests that k-slow burning combines difficulties of both graph burning and k-broadcasting. As k-slow burning and graph burning coincide for $k \geq \Delta(G)$ (the maximum degree of G), this variant generalises normal graph burning.

Our Results. The aim for this work is to contrast k-slow burning (for some fixed k) with graph burning, focusing on two main aspects of the problem, complexity and upper bounds, in Sects. 2 and 3, respectively. Due to the page limit, some of the proofs will be omitted.

Complexity. Graph burning is known to be $\mathcal{NP}$-hard, even for very simple graph classes such as path forests and spider graphs [1]. To match both of these results for k-slow burning, we will use a very similar reduction as for normal graph burning to show hardness for path forests, before then connecting this path forests components in a suitable way to show $\mathcal{NP}$-hardness for spiders. By also showing $\mathcal{NP}$-hardness of the problem when restricted to graphs of radius 1, we will find a class of graphs on which k-slow burning is hard, whereas normal graph burning is solvable in polynomial time. Finally, we will see that even the subproblem of checking a potential k-slow burning sequence for correctness, while trivial for normal graph burning, is still $\mathcal{NP}$-hard. The latter two results verify our observation in the preceding paragraph that k-slow burning seems to be harder than normal graph burning.

Upper Bounds. In an attempt to find an analogy for the burning number conjecture in regard to k-slow burning, we study upper bounds for the k-slow burning number on connected graphs. Here, the k-slow burning number of the star graph S_{n-1} on n vertices, $b_s(k, S_{n-1})$, plays a critical role for discussing upper bounds. For $k = 1, 2$ we will show, that indeed $b_s(k, G) \leq b_s(k, S_{n-1})$ holds for all connected graphs G with n vertices. While this is no longer true for $k \geq 3$, we can still show $b_s(k, G) \leq b_s(k, S_{n-1}) + f(k)$ for some error term $f(k)$ that does not depend on n and that satisfies $f(k) = \Theta(k)$.

2 Complexity

In this section, we will prove $\mathcal{NP}$-hardness of k-slow burning on several different graph classes. We will start by showing hardness on path forests, which is known to be $\mathcal{NP}$-hard for normal graph burning from [1]. This result immediately implies the $\mathcal{NP}$-hardness of k-slow burning for $k \geq 2$ on path forests, as $k \geq \Delta(G)$. For $k = 1$, we will use an analogous reduction from a variant of the 3-Partition problem as for normal graph burning. This variant is known to be strongly $\mathcal{NP}$-hard due to [13].

DISTINCT 3- PARTITION (D3P)
Input: A Set $A = \{a_1, ..., a_{3n}\}$ of pairwise distinct integers $a_i \in \mathbb{N}$ and a natural number B s.t. $n \cdot B = \sum_{i=1}^{3n} a_i$ and $B/4 < a_i < B/2$ for all i.
Question: Can A be partitioned into triples $(b_1^{(i)}, b_2^{(i)}, b_3^{(i)})$, $i = 1, ..., n$ s.t. $b_1^{(i)} + b_2^{(i)} + b_3^{(i)} = B$ for all i?

In our reduction we will need the 1-slow burning number of the path P_n.

Lemma 1 *Let P_n be the path on n vertices. We have* $b_s(1, P_n) = \lceil \sqrt{n - 3/4} + 1/2 \rceil$.

The proof is analogous to the one for normal graph burning in [4].

Theorem 1 *1-slow burning is $\mathcal{NP}$-hard, even when restricted to path forests.*

Proof Let $(A := \{a_1, \ldots, a_{3n}\}, B)$ be an instance of D3P. W.l.o.g., we assume $B = O(\text{poly}(n))$, which is possible since D3P is strongly $\mathcal{NP}$-hard. Also, for sufficiently large values of n we have $1 \notin A$. We define $M := \max A$ and $Y := \{2, 3, \ldots, M\} \setminus A$. We construct a path forest P consisting of n paths $P^{(i)}_{2B-6}$ of length $2B-6$, one path P_{2y-2} of length $2y-2$ for each $y \in Y$ as well as one additional isolated vertex. We will prove that the D3P-instance is solvable iff $b_s(1, P) \le M$.

First, assume that the instance of D3P is solvable and thus there exists a partition of A into n triples $(b^{(i)}_1, b^{(i)}_2, b^{(i)}_3)$, $i = 1, \ldots, n$, that all sum to B. Then we can cover $P^{(i)}_{2B-6}$ with 3 paths of lengths $2b^{(i)}_1 - 2$, $2b^{(i)}_2 - 2$ and $2b^{(i)}_3 - 2$, i.e., with burning neighbourhoods of sizes $b^{(i)}_j$. Doing so, only the paths of length $2y-2$, $y \in Y$ and the isolated vertex remain. These paths correspond to burning neighbourhoods of sizes $y \in Y$ and 1 respectively. As $A \cap Y = \emptyset$ and $1 \notin A$ we used every burning range exactly once and thus $b_s(1, P) \le M$.

Conversely, assume that P can be burned in at most M rounds. Note, that in this case we already have $b_s(1, P) = M$ as P is a subgraph of $P_{|V(P)|}$ with

$$|V(P)| = (2B-6)n + \left(\sum_{y \in Y} 2y - 2\right) + 1 = \left(\sum_{i=1}^{3n} 2a_i - 2\right) + \left(\sum_{y \in Y} 2y - 2\right) + 1$$
$$= \left(\sum_{i=2}^{M} 2i - 2\right) + 1 = M^2 - M + 1$$

and thus $b_s(1, P) \ge b_s(1, P_{M^2-M+1}) = M$ by Lemma 1. Also, all burning neighbourhoods have to be disjoint, since otherwise we can only burn less than $M^2 - M + 1 = |V(P)|$ vertices.

Next, note that in order to burn a path of length $2B-6$, we need to use at least 3 burning ranges as otherwise we can burn at most $(2b_1 - 2) + (2b_2 - 2) < (2(B/2) - 2) + (2((B/2) - 1) - 2) = 2B - 6$ vertices. Since there are $|Y| + 1$ other paths, each needing at least one burning range, we need a minimum of $3n + |Y| + 1 = M$ burning ranges in total. This also implies, that we have to use exactly 3 burning ranges for each path of length $2B-6$ and exactly one burning range for every other path. This means, that we have to use the burning range y to cover the path of length $2y-2$ for each $y \in Y$ and burning range 1 for P_1. This way, only the burning ranges $a_1, \ldots, a_{3n}$ remain to cover the n paths of length $2B-6$. This induces a partition of A into triples by choosing $(b^{(i)}_1, b^{(i)}_2, b^{(i)}_3)$ as exactly the 3 burning ranges used to cover $P^{(i)}_{2B-6}$. □

This reduction is very similar to the one used for normal graph burning in [1]. Here, P consists of n paths of length $2B-3$ and one path each of length $2y-1$, $y \in Y := \{1, \ldots, M\} \setminus A$. Note, that if the constructed instance of (k-slow) burning (P, t) is solvable, we always have $b_s(1, P) = M$ (and never $b_s(1, P) < M$).

2.1 Connecting the Paths

To show $\mathcal{NP}$-hardness on spider graphs we will use a similar approach as in [1] and connect the components of the path forest P from Theorem 1 to a spider graph S with central vertex v in a way, which ensures that in order to burn S in $t+1$ steps (a) we have to choose v as our first source of fire and (b) upon removing the burning neighbourhood of v only a subgraph of the path forest P remains. Together with Theorem 1 this proves $\mathcal{NP}$-hardness on spider graphs.

Theorem 2 *Let $k \geq 1$ be fixed. k-slow burning is $\mathcal{NP}$-hard, even when restricted to spider graphs.*

Proof For this proof, we will identify path forests and spider graphs G with the list of lengths of their paths (legs), writing $G = [\ell_1, \ldots, \ell_m]$. Let $(P = [p_1, \ldots, p_m], t)$ be the instance of k-slow burning on path forests constructed in the reduction for Theorem 1 for $k = 1$ or in [1] for $k \geq 2$ respectively. We construct the spider graph (sketched in Fig. 1 for $k = 1$)

$$S = [p_1, \ldots, p_m, \underbrace{1, \ldots, 1}_{k}, \underbrace{2, \ldots, 2}_{k}, \ldots, \underbrace{t, \ldots, t}_{k}]$$

and show that $b_s(k, S) \leq t + 1$ iff $b_s(k, P) \leq t$.

By v, we denote the central vertex of S. We will show that S can never be burned in time if v is not part of the burning neighbourhood of the first source of fire. For this, note that

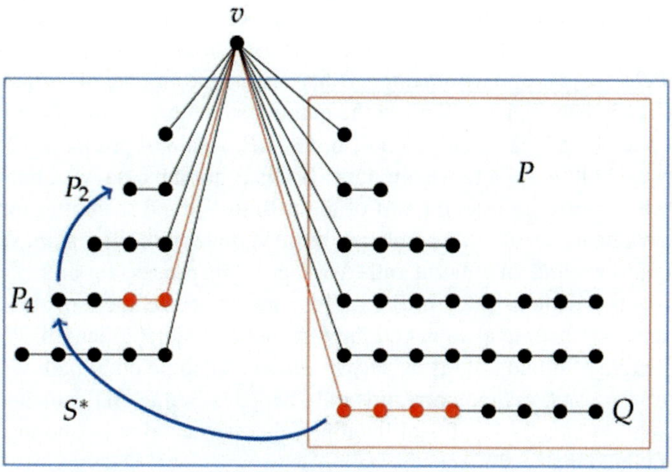

Fig. 1 U_v is marked red. We search for a set of paths to embed them into $Q \in P$. Starting from v, 4 vertices in Q have been burned. Thus, we look at a path of length 4 in $S^* - P$. This path contains 2 burned vertices, so we consider a path P_2. As this path is completely unburnt, we can embed P_2, as well as the unburnt parts of P_4 and Q in $S - U_v$ into Q

$$|V(S)| = |V(P)| + 1 + \left(k \sum_{j=1}^{t} j\right) = \begin{cases} (1 + \sum_{j=2}^{t} 2j - 2) + \sum_{j=1}^{t} j, & k = 1, \\ t^2 + 1 + k \sum_{j=1}^{t} j, & k \geq 2. \end{cases} \tag{1}$$

In the burning neighbourhood containing v we can ignite one vertex in the first step and $k(t^* - 1)$ vertices in step $t^* \geq 2$. Every other burning neighbourhood of size t^* is a path and thus we can burn 1 vertex for $k = 1$ and $t^* = 1$, at most $2t^* - 2$ vertices for $k = 1$ and $t^* \geq 2$ and at most $2t^* - 1$ vertices for $k \geq 2$. Hence, if v lies in the burning neighbourhood of size t^*, we can burn at most

$$\begin{cases} \left(1 + \sum_{j=2}^{t+1}(2j - 2)\right) - (2t^* - 2) + \left(1 + \sum_{j=1}^{t^*-1} j\right), & k = 1 \\ (t + 1)^2 - (2t^* - 1) + \left(1 + k \sum_{j=1}^{t^*-1}\right), & k \geq 2 \end{cases} \tag{2}$$

vertices in total. This is equal to $|V(S)|$ iff $t^* = t + 1$. Also, to achieve equality between (1) and (2), the number of vertices in each burning neighbourhood has to be maximal, so the burning neighbourhood of size $t + 1$ containing v has to be of size $1 + k \sum_{j=1}^{t} j$. To achieve this we have to choose v as our first source of fire for $k \geq 2$ or v has to be one of the first two burning vertices for $k = 1$. In the latter case, w.l.o.g. we can also assume v to be the first source of fire.

Thus, we burn v as our first source of fire and fix some arbitrary, maximal, burning neighbourhood U_v (of size $t + 1$) of v. Consider the path forest P^* induced by $S - U_v$ as a subgraph of S and denote by S^* the path forest that remains after deleting v from S. Because we choose U_v to be maximal, we have $|P^*| = |P|$. We claim, that P^* can always be embedded as a subgraph into P and will construct this embedding algorithmically. To do so, we look at every path p^* of length l^* in P separately and find a set of unmarked paths $S_{p^*} = \{p_1, \ldots, p_m\}$ of lengths $\{l_1, \ldots, l_m\}$ in P^* such that $l_1 + \cdots + l_m = l^*$. After this, we mark the paths $p_1, \ldots, p_m$. If p^* is part of P^*, we choose $S_{p^*} = \{p^*\}$ and we are done. Otherwise, some number of vertices n_1 out of p^* have already been burned. We add the unburnt subpath of p^* to S_{p^*} and continue to look at one of the k paths of length n_1 in $S^* - P$. Again, if this path is completely unburnt, we add it to S_{p^*} and we are done. If this is not the case, we continue in the same way by adding the unburnt part of the path to S_{p^*} and searching for a path of length n_2, where n_2 denotes the number of burnt vertices in the new path. We proceed to do this until we find an unburnt path. We repeat this process for all paths $p^* \in P$. This approach is demonstrated in Fig. 1. To prove the correctness of this approach, we have to show that we always find enough paths of suitable length. We look for a path of length l^* if and only if exactly l^* vertices of some other path were burned starting from v. As U_v is maximal, this only happens k times. This means, because $S^* - P$ contains exactly k paths of length l^*, that we can always find an unmarked path of suitable length.

This proves the claim and therefore we have $b_s(k, P^*) \geq b_s(k, P)$, which means that it is optimal to choose U_v in a way such that $S - U_v = P$. This means that $b_s(k, S) = t + 1$ iff $b_s(k, P) = t$, which concludes the proof. □

2.2 k-Slow Burning is Harder Than Graph Burning

So far we have only considered graph classes where normal graph burning is known to be $\mathcal{NP}$-hard. We want to find a class of graphs, where k-slow burning is $\mathcal{NP}$-hard while normal graph burning is solvable in polynomial time. For this, we will look at classes C, where the radius of all graphs is bounded by some constant c. In this case we have $b(G) \leq c + 1$ for all $G \in C$ and thus we can simply check all burning sequences of length at most $c + 1$. This is easily possible in polynomial time. In contrast to that, k-slow burning is hard, even when restricted to graphs of radius 1. To prove this, we will use a reduction from the k-broadcasting problem.

MINIMUM TIME k- BROADCASTING
Input: A Graph G, an integer k and a time bound t.
Question: Is a k-broadcast on G starting from an optimal source of information v possible in t time steps, i.e., does $\text{minb}_k(G) := \min_{v \in V} b_k(v, G) \leq t$ hold?

Note that the source of information is already informed before the first step of the broadcasting process, whereas for k-slow burning the first source of fire is ignited during the first time step. For $k = 1$ this problem is $\mathcal{NP}$-hard according to [8]. The reduction can easily be adjusted to also hold for $k \geq 2$ and to show hardness on planar graphs.

Theorem 3 *Let $k \geq 1$ be fixed. k-slow burning is $\mathcal{NP}$-hard, even when restricted to graphs of radius 1.*

Proof Let (G, t) be an instance of k-broadcasting. We extend G by adding an independent set $I = \{w_1, \ldots, w_{|I|}\}$ of size $(t + 1)(k + 1) - 1$ to G and connect all vertices in I as well as in $V(G)$ to some new vertex v. The resulting graph G^* clearly has radius 1. We show that $\text{minb}_k(G) \leq t$ holds iff $b_s(k, G^*) \leq t + 2$.

First, assume that $\text{minb}_k(G) \leq t$ holds with $\text{minb}_k(G) = b_k(x, G)$ for some vertex $x \in V(G)$. Then, $(v, w_1, \ldots, w_{t+1})$ forms a k-slow burning sequence of length $t + 2$ for G^*. To burn down G^* in $t + 2$ steps, we spread the fire from v to x as well as $k - 1$ vertices in I in round 2 and to k vertices in I in each subsequent round. This way we ignite exactly $(k - 1) + tk$ vertices in I starting from v and choose $t + 1$ vertices in I as sources of fire. This means, because of $(k - 1) + (t + 1) + tk = |I|$, that I is burned in time. Also, starting from round 3, we can replicate the behaviour from the k-broadcast on G to ignite every vertex in $V(G)$ by round $t + 2$.

Conversely, assume that $b_s(k, G) \leq t + 2$. Clearly, we can assume that the first source of fire does not lie in I. Out of the $(t + 1)(k + 1) - 1$ vertices in I we can ignite a maximum of $k + 1$ each round - one as a source of fire and k coming from v. As we have to ignite all $(t + 1)(k + 1) - 1$ vertices within rounds $2, \ldots, t + 2$, we have to ignite at least k vertices in I in round 2. Thus, v has to be the first source of fire and in rounds $2, \ldots, t + 2$ we can ignite only one vertex x in $V(G)$ coming from v. By assumption, we have $b_s(k, G) \leq t + 2$, so, if the fire spreads to x in round t^*, x can ignite G within rounds $t^* + 1, \ldots, t + 2$. If we copy this behaviour as a broadcast-protocol on G starting from x, we get $\text{minb}_k(G) \leq b_k(x, G) \leq (t + 2) - (t^* + 1) + 1 \leq t$. □

2.3 Checking Burning Sequences

Whereas checking whether a given sequence $(v_1, \ldots, v_t)$ works as a burning sequence for a given graph is trivial for normal graph burning, this is not the case for k-slow burning. Using a reduction from k-broadcasting we will show that this subproblem is in fact $\mathcal{NP}$-hard for arbitrary graphs.

Theorem 4 *Given a fixed integer $k \geq 1$, a graph G and a sequence of vertices $(v_1, \ldots, v_t)$, it is $\mathcal{NP}$-hard to decide, whether $(v_1, \ldots, v_t)$ is a suitable burning sequence for G, even when restricted to planar graphs. However, this problem is solvable in polynomial time when restricted to trees.*

Proof Let (G, t, v_1) be an instance of k-broadcasting for some planar graph G. We extend G by adding t isolated vertices $\{v_2, \ldots, v_{t+1}\}$ and denote the resulting graph by G^*. Clearly, G^* is planar and a k-broadcast on G starting from v_1 in t time steps is possible iff $(v_1, \ldots, v_{t+1})$ is a possible burning sequence for G^*. This proves the $\mathcal{NP}$-hardness on planar graphs. □

In contrast to this result, when restricting ourselves to trees, checking whether a potential k-slow burning sequence is feasible is possible in polynomial time by using a polynomial time algorithm for k-broadcasting on trees [10]. The exact algorithm and proof of correctness will be omitted here.

3 Upper Bounds

The burning number conjecture, stating $b(G) \leq \lceil \sqrt{n} \rceil$ for all connected graphs G on n vertices, is the central open question within the field of graph burning, with numerous papers either proving the conjecture for certain classes of graphs (such as Hamiltonian graphs, 2-caterpillars or spiders) or proving upper bounds close to $\lceil \sqrt{n} \rceil$. In this section, we will look at the question of finding upper bounds for $b_s(k, G)$ for fixed values of k. Similar to normal graph burning, for this purpose we only need to consider trees, as

$$b_s(k, G) = \min\{b_s(k, T) \mid T \text{ spanning tree of } G\}.$$

When discussing upper bounds for $b_s(k, G)$, the star graph S_{n-1} on n vertices fulfills a critical role. To burn down S_{n-1}, it is clear that it is optimal to choose the central vertex v as the first source of fire. In every subsequent step, except for the last step, if there are less than $k + 1$ non-burning vertices left, the fire spreads to exactly $k + 1$ of the outer vertices: k vertices can be ignited by v and one additional vertex can be chosen as a new source of fire. Using this observation we get $b_s(k, S_n) = \lceil (n + k)/(k + 1) \rceil$. Seeing that on S_n the number of vertices burned largely remains constant during each step of the burning process, it seems reasonable to assume that $b_s(k, S_n)$ is close to the upper bound for $b_s(k, G)$ among all connected graphs G, at least for large values for n.

The cases $k = 1$ and $k = 2$. During the 1- or 2-slow burning process on S_n, we can burn exactly 2 or 3 vertices in each step of the burning process, respectively. If for an arbitrary tree T we can find 2 (respectively, 3) vertices to be burned in every step of the process, this proves that $b_s(k, S_n)$ indeed marks an upper bound for the 1- and 2-slow burning number on connected graphs.

Theorem 5 *Let G be an arbitrary connected graph with n vertices. For $k \in \{1, 2\}$ we have*

$$b_s(k, G) \leq b_s(k, S_n) = \left\lceil \frac{n+k}{k+1} \right\rceil.$$

Proof Let T be a tree, $k \in \{1, 2\}$ and $t_{max} = \lceil (n+k)/(k+1) \rceil$. By $BN_{t^*}(v_1)$ we denote the set of all vertices burned starting from the first source of fire v_1 after t^* steps. It suffices to show that we can choose v_1 and $BN_{t_{max}}(v_1)$ in a way such that $|V(G) \setminus BN_{t_{max}}(v_1)| \leq t_{max} - 1$.

Case $k = 1$: Choose v_1 arbitrarily. In each of the time steps $t^* \in \{2, \ldots, t_{max}\}$ we can extend the subtree formed by $BN_{t^*-1}(v_1)$ by one arbitrary vertex. Thus, we have $|BN_{t_{max}}(v_1)| = t_{max}$ which leads to

$$n - |BN_{t_{max}}(v_1)| = n - \left\lceil \frac{n+1}{2} \right\rceil = \left\lfloor \frac{n-1}{2} \right\rfloor \leq t_{max} - 1.$$

Case $k = 2$: We choose v_1 to be a 1/2-separator of T, i.e., in a way, that each of the components $C_1, \ldots, C_m$ in $T - \{v_1\}$ satisfies $|C_i| \leq |T|/2$. Using a greedy approach, we can partition the indices $1, \ldots, m$ into two sets I_1 and I_2 in a way, that we have

$$\frac{n-1}{3}|T| \leq \sum_{i \in I_j} |C_i| \leq \frac{2(n-1)}{3}|T|, \quad j = 1, 2.$$

By T_j, $j = 1, 2$, we denote the subtree of T induced by $\{v_1\} \cup \bigcup_{i \in I_j} C_i$. This way we have $|T_j| \geq t_{max}$, $j = 1, 2$, so in every subsequent step of the burning process $t^* \in \{2, \ldots, t_{max}\}$ we can extend $BN_{t^*}(v_1) \cap T_j$, $j = 1, 2$ by one vertex each and thus we get $|BN_{t_{max}}(v_1) \cap T_j| = t_{max}$ which (because of $T_1 \cap T_2 = \{v_1\}$) leads to $|BN_{t_{max}}| = 2t_{max} - 1$. If we choose $B_{t_{max}}(v_1)$ in this way, we get

$$n - |BN_{t_{max}}(v_1)| = n - \left(2\left\lceil \frac{n+2}{3} \right\rceil - 1\right) \leq \left\lceil \frac{n+2}{3} \right\rceil - 1.$$

The last inequality holds because of $3\lceil (n+2)/3 \rceil - 2 \geq (3n+6)/3 - 2 = n$. □

Asymptotic Upper Bounds for $k \geq 3$. For larger values of k, the burning number of the star graph is no longer always the worst case. For example, taking $k = 3$ and $n = 5$ we get $b_s(3, S_4) = 2 < 3 = b_s(3, P_5)$. However, we can show, that after some *preparation phase* consisting of at most $f(k)$ rounds (for some suitable function f) we can burn at least $k + 1$ vertices in each subsequent step. Thus, for fixed k, $b_s(k, S_{n-1})$ forms an asymptotic bound for $b_s(k, G)$ on all connected graphs G.

Theorem 6 *Let $k \geq 3$ be fixed and G be a connected graph with n vertices. Then,*

$$b_s(k, G) \leq \left\lceil \frac{n}{k+1} \right\rceil + f(k)$$

holds for some function f satisfying $f(k) \leq 2k - 3$.

Proof We assume $k < n$ as otherwise the theorem is trivial. To prove the bound for $f(k)$ for an arbitrary tree T, we use a similar approach as for the case $k = 2$. First, in a *preparation phase*, we iteratively choose 1/2-separators S in T in a way ensuring that all components in the forest $T - S$ are sufficiently small. In the first $|S|$ rounds of the burning process, we only ignite all vertices in S. To obtain an upper bound for $|S|$ we make use of the following lemma. □

Lemma 2 *Let $1 \leq Z \in \mathbb{R}$ be arbitrary and let T be a tree. We choose $c \in \mathbb{N}_0$ in a way such that $cZ \leq |V(T)| \leq (c+1)Z$ holds. Then there exists some set $S \subset V(T)$ such that each component in $T - S$ has at most Z vertices and such that $|S| = 0$ for $c = 0$ and $|S| \leq 2c - 1$ otherwise.*

The proof of this lemma uses a simple induction over c and will be omitted here. By using the lemma with $Z = n/k \geq 1$ we obtain a set of vertices S such that $|S| \leq 2k - 3$ and every component C_i in the forest $F := T - S$ contains at most n/k vertices. In the first $|S|$ rounds of the burning process we only ignite all vertices in S. We denote the components of $T - S$ by $C_1, \ldots, C_m$, their orders by $n_i := |C_i|$ and we assume $\lfloor n/k \rfloor \geq n_1 \geq \ldots \geq n_m$. Furthermore, in each of the components we fix some root r_i that is adjacent to at least one vertex in S. We have to burn $T - S$ in $\lceil n/(k+1) \rceil$ rounds.

To do so, we fix the set of remaining sources of fire, $M = \{w_1, \ldots, w_{\lceil n/(k+1) \rceil}\}$, in a way that ensures every component in $(T - S) - M$ contains at most $\lceil n/(k+1) \rceil$ vertices. In $T - S$ there are at most k components C_i that contain more than $n/(k+1)$ vertices, as otherwise $T - S$ would have to contain at least $(k+1)((n/(k+1)) + 1) > n$ vertices in total. Thus, we have to remove at most $k[(n/k) - (n/(k+1))] \leq \lceil n/(k+1) \rceil$ vertices, which ensures the existence of a suitable set of sources of fire M. We choose the vertices in M in a way that (a) all components in $T - S$ remain connected and (b) the root r_i of every component in $T - S$ is not chosen as a source of fire as long as there are still other unburnt vertices in C_i.

We denote the remaining forest $T - S - M$ as $F_{\lceil n/(k+1) \rceil}$. For descending $j = \lceil n/(k+1) \rceil, \ldots, 1$ we construct the graph F_{j-1} by removing one vertex in each of the k largest components of F_j or one vertex out of every component if there are at most k components remaining. Via induction over $j = \lceil n/(k+1) \rceil, \ldots, 0$, we show that F_j satisfies the following two properties:

1. Every component in F_j contains at most j vertices.
2. Either it was possible to burn k vertices in every round of the burning process, starting from round $|S| + 1$ *or* F_j contains at most k components.

As this means F_0 is the empty graph and because we can clearly obtain F_j from F_{j+1} within one round of the burning process, this will prove the upper bound for $b_s(k, T)$. We start by looking at the base case $j = \lceil n/(k+1) \rceil$. We have shown (1) previously and for (2) there is nothing to show as we are exactly in round $|S| + 1$ of the burning process. Now, assume $j < \lceil n/(k+1) \rceil$ and that both properties have been shown to hold for larger j.

By this assumption, in F_{j+1} every component contains at most $j + 1$ vertices. If F_{j+1} contains at most k components we can burn one vertex in each component and the remaining graph still has at most k components containing $\leq j$ vertices each. Thus, (1) and (2) both hold. Otherwise, by induction, it was possible to remove k vertices staring from S in each of the previous rounds $|S| + 1, \ldots, |S| + \lceil n/(k + 1) \rceil - j$. F_{j+1} cannot contain more than k components with exactly $j + 1$ vertices, as otherwise $T - S - M$ would have to contain

$$(k+1)(j+1) + k\left(\left\lceil \frac{n}{k+1} \right\rceil - j\right) \geq k + j + 1 + \frac{kn}{k+1} > n - \left\lceil \frac{n}{k+1} \right\rceil \geq |T - S - M|$$

vertices, which is a contradiction. Hence, after igniting one vertex each in the k largest components of F_{j+1}, every component in the resulting forest F_j contains at most k vertices. Thus, (1) holds. As we removed k vertices in order to construct F_j from F_{j+1}, (2) is also true. This proves the theorem.

We compare the k-slow burning number of paths and star graphs on n vertices for $k \geq 3$ and to do so maximise $b_s(k, P_n) - b_s(k, S_{n-1}) = \sqrt{n} - (n/(k+1)) + O(1)$. The function $g_k(n) := \sqrt{n} - (n/(k+1))$ reaches its maximum for $n = ((k + 1)/2)^2$. In this case we have $g_k(n) = (k+1)/4 = \Omega(n)$. This means that the bound for $f(k)$ from Theorem 6 already has the correct magnitude, i.e., we necessarily have $f(k) = \Theta(k)$.

4 Concluding Remarks

We have proven $\mathcal{NP}$-hardness on path forests, spider graphs and graphs of radius 1. Also, an analogous approach as in [12] can be used to show hardness on caterpillars of maximum degree 3 as well. Out of our three results, the last is the most notable, as for the class of graphs of radius 1 normal graph burning is possible in polynomial time. Together with our observation that checking a potential k-slow burning sequence for correctness is hard, this indicates that k-slow burning is in fact harder than normal graph burning. Natural follow-up questions would be to discuss the complexity on other graph classes where normal graph burning is known to be easy, such as split graphs and cographs, or to discuss the parameterised complexity of the problem.

In an attempt to find an analogy to the well known burning number conjecture, we then studied upper bounds for the k-slow burning number. Whereas the conjectured upper bound of $\lceil \sqrt{n} \rceil$ holds with equality for paths for normal graph burning, for k-slow burning the star graph S_{n-1} appears to be critical for studying upper bounds.

We could prove, that $b_s(k, G)$ is indeed maximal on star graphs for $k \in \{1, 2\}$ while for $k \geq 3$ this is only the case asymptotically.

These observations lead to the question whether there is some k and a connected graph G^* on n vertices, such that $b_s(k, G^*) > \max\{b_s(k, P_n), b_s(k, S_{n-1})\}$. We conjecture that this is not the case which means that we have

$$b_s(k, G) \leq \max\{b_s(k, P_n), b_s(k, S_{n-1})\}$$

for all connected graphs G. For $k \in \{1, 2\}$ we have proven this conjecture to hold, while for $k \geq 3$ it remains open. If we assume $n \geq 2$ and $k \geq n$, we have $b_s(k, S_{n-1}) = 2$ and thus $\max\{b_s(k, S_{n-1}), b_s(k, P_n)\} = b_s(k, P_n) = \lceil\sqrt{n}\rceil$. Therefore the proposed k-slow burning conjecture generalises the original conjecture. Although, because of this, proving the conjecture seems difficult, it may be interesting to show the conjecture to hold on certain subclasses of trees such as caterpillars or spiders where the normal burning conjecture is known to be true.

References

1. Bessy, S., Bonato, A., Janssen, J., Rautenbach, D., Roshanbin, E.: Burning a graph is hard. Discret. Appl. Math. **232**, 73–87 (2017). https://doi.org/10.1016/j.dam.2017.07.016
2. Bonato, A.: A survey of graph burning. Contributions Discret. Math. **16**, 185–197 (2020). https://doi.org/10.11575/CDM.V16I1.71194
3. Bonato, A., Janssen, J., Roshanbin, E.: Burning a graph as a model of social contagion. In: Bonato, A., Graham, F., Prałat, P. (eds) Algorithms and Models for the Web Graph, LNCS, vol. 8882 (2014). https://doi.org/10.1007/978-3-319-13123-8_2
4. Bonato, A., Janssen, J., Roshanbin, E.: How to burn a graph (2015). https://doi.org/10.48550/arXiv.1507.06524
5. Bonato, A., Kamali, S.: Approximation algorithms for graph burning (2018). https://doi.org/10.48550/arXiv.1811.04449
6. Bonato, A., Kamali, S.: An improved bound on the burning number of graphs (2021). https://doi.org/10.48550/arXiv.2110.01087
7. Bonato, A., Lidbetter, T.: Bounds on the burning numbers of spiders and path-forests. Theor. Comput. Sci. **794**, 12–19 (2019). https://doi.org/10.1016/j.tcs.2018.05.035
8. Garey, M., Johnson, D.: Computers and Intractability. A Guide to the Theory of NP-Completeness. W. H. Freeman & Co., New York (1990)
9. Gautam, R., Kare, A., Durga, B.: Improved approximation algorithm for graph burning on trees (2022). https://doi.org/10.48550/arXiv.2204.00772
10. Harutyunyan, H., Liestmann, A., Shao, B.: A linear algorithm for finding the k-broadcast center of a tree. Networks **53**, 287–292 (2009). https://doi.org/10.1002/net.20270
11. Hedetniemi, S.M., Hedetniemi, S.T., Liestmann, A.: A survey of gossiping and broadcasting in communication networks. Networks **18**, 319–349 (1988)
12. Hiller, M., Triesch, E., Koster, A.: On the burning number of p-Caterpillars. In: Gentile, C., Stecca, G., Ventura, P. (eds) Graphs and Combinatorial Optimization: from Theory to Applications. AIRO Springer Series, vol. 5 (2021)
13. Hulett, H., Will, T., Woeginger, G.: Multigraph realizations of degree sequences: maximization is easy, minimization is hard. Oper. Res. Lett. **36**, 594–596 (2008)
14. Moghbel, D.: Topics in graph burning and datalog. Thesis (2022). https://doi.org/10.32920/19775380.v1

15. Norin, S., Turcotte, J.: The burning number conjecture holds asymptotically (2022). https://doi.org/10.48550/arXiv.2207.04035
16. Slater, P.J., Cockayne, E.J., Hedetniemi, S.T.: Information dissemination in trees. SIAM J. Comput. **10**, 692–701 (1981)

Discrepancies of Subtrees

Tarun Krishna, Peleg Michaeli, Michail Sarantis, Fenglin Wang, and Yiqing Wang

Abstract We study multicolour, oriented and high-dimensional discrepancies of the set of all subtrees of a tree. As our main result, we show that the r-colour discrepancy of the subtrees of any tree is a linear function of the number of leaves ℓ of that tree. More concretely, we show that it is bounded by $\lceil (r-1)\ell/r \rceil$ from below and $\lceil (r-1)\ell/2 \rceil$ from above, and that these bounds are asymptotically sharp. Motivated by this result, we introduce natural notions of oriented and high-dimensional discrepancies and prove bounds for the corresponding discrepancies of the set of all subtrees of a given tree as functions of its number of leaves.

1 Introduction

Given a hypergraph $\mathcal{H} = (\mathcal{V}, \mathcal{E})$, a (two-)colouring of (the vertices of) $\mathcal{H}$ is a function $f : \mathcal{V} \to \{\pm 1\}$. For a hyperedge A we set $f(A) = \sum_{a \in A} f(a)$, and $|f(A)|$ is called the **imbalance** of A. The (combinatorial) **discrepancy** of $\mathcal{H}$ is defined to be

$$\mathcal{D}(\mathcal{H}) = \min_{f:\mathcal{V}\to\{\pm 1\}} \max_{A\in\mathcal{E}} |f(A)|.$$

T. Krishna · P. Michaeli (✉) · M. Sarantis · F. Wang · Y. Wang
Department of Mathematical Sciences, Carnegie Mellon University, Pittsburgh, PA 15213, USA
e-mail: pelegm@cmu.edu; pmichael@andrew.cmu.edu

T. Krishna
e-mail: tkrishna@andrew.cmu.edu

M. Sarantis
e-mail: msaranti@andrew.cmu.edu

F. Wang
e-mail: fenglinw@andrew.cmu.edu

Y. Wang
e-mail: yiqingw@andrew.cmu.edu

A. Brieden et al. (eds.), *Graphs and Combinatorial Optimization: from Theory to Applications*, AIRO Springer Series 13,
https://doi.org/10.1007/978-3-031-46826-1_7

Namely, the discrepancy of $\mathcal{H}$ is the maximum imbalance of an edge under an optimal colouring. It is often convenient to think about this definition in terms of a game: an adversary colours $\mathcal{V}$ using 2 colours. He tries to do it as balanced as possible, that is, so that the distribution of the colours in every member of $\mathcal{E}$ will be as close as possible to uniform. Our goal is then to find a member of $\mathcal{E}$ of maximum imbalance. Over the last century, the study of discrepancy-type problems has developed into a field with extensive range and variety, demonstrating strong ties to number theory, Ramsey theory, and computational methods. We refer the reader to the book of Matoušek [13] for a comprehensive overview of the topic.

There are several natural ways to generalise the above definition of (2-colour) discrepancy to an arbitrary number of colours. One such generalisation was introduced by Doerr and Srivastav [4], in which the notion of imbalance captures the maximum deviation of the size of a colour class from the mean size of a colour class (or, in other words, the (scaled) ℓ^∞-distance of the colour distribution from the uniform distribution). We call it here the **symmetric r-colour discrepancy** of $\mathcal{H}$, and denote[1]

$$\mathcal{D}_r^\circ(\mathcal{H}) = \min_{f:\mathcal{V}\to[r]} \max_{A\in\mathcal{E}} \max_{j\in[r]} \left| r\left|f^{-1}(j)\cap A\right| - |A|\right| .$$

Recently, mostly in the context of graphs, a slightly different notion of multicolour discrepancy was studied, in which the notion of imbalance captures the deviation of the size of the *largest* colour class from the mean size. We call it here the (upper) **r-colour discrepancy** of $\mathcal{H}$, and denote

$$\mathcal{D}_r(\mathcal{H}) = \min_{f:\mathcal{V}\to[r]} \max_{A\in\mathcal{E}} \max_{j\in[r]} \left(r\left|f^{-1}(j)\cap A\right| - |A|\right) .$$

It is not hard to see that these definitions are both generalisations of the classical notion of discrepancy, and differ from each other by a constant factor. Concretely, $\mathcal{D}_2(\mathcal{H}) = \mathcal{D}_2^\circ(\mathcal{H}) = \mathcal{D}(\mathcal{H})$ and

$$\mathcal{D}_r(\mathcal{H}) \le \mathcal{D}_r^\circ(\mathcal{H}) \le (r-1)\mathcal{D}_r(\mathcal{H}) \tag{1}$$

for every hypergraph $\mathcal{H}$ and $r \ge 2$.

The "upper" variation is more natural in the context of edge-colourings in (hyper)graphs, due to its direct relation to Ramsey-type questions: given an edge-colouring of a graph, instead of looking for a monochromatic copy of a target subgraph, one looks for a copy of that subgraph in which one of the colours appears (significantly) more than the average. In that sense, discrepancy-type problems may be considered as a relaxation—or rather a quantification—of Ramsey-type problems.

Let us elaborate on combinatorial discrepancies in the context of graphs. Here, given a base graph G and a family of graphs $\mathcal{X}$, we construct a hypergraph whose vertices are the edges of G and whose hyperedges are edge sets that from a member

[1] The original definition of Doerr and Srivastava was a $(1/r)$-scaling of the above definition; we scaled it for convenience to allow $\mathcal{D}_2^\circ = \mathcal{D}$, and to ensure it is an integer.

of $\mathcal{X}$. The **discrepancy of $\mathcal{X}$ in G**, denoted $\mathcal{D}(G, \mathcal{X})$, is the discrepancy of that hypergraph. Analogously, we define the symmetric r-colour discrepancy of $\mathcal{X}$ in G ($\mathcal{D}_r^\circ(G, \mathcal{X})$) and the r-colour discrepancy of $\mathcal{X}$ in G ($\mathcal{D}_r(G, \mathcal{X})$). It is helpful to keep in mind that $\mathcal{D}(\mathcal{H})$ is monotone in $\mathcal{E}$, hence $\mathcal{D}(G, \mathcal{X})$ is monotone both in G and in $\mathcal{X}$. It is therefore natural (and often nontrivial) to study $\mathcal{D}(K_n, \mathcal{X})$.

The study of combinatorial discrepancy in graphs was initiated by Erdős et al. [5], who analysed the 2-colour discrepancy of a fixed spanning tree with a given maximum degree in the complete graph. However, several earlier results can be stated using this terminology. As an example we mention the result of Erdős and Spencer [6], that can be interpreted as showing that the (2-colour) discrepancy of cliques in the complete graph on n vertices (or, more generally, of hypercliques in the complete k-uniform hypergraph) is of order $n^{3/2}$ (or, more generally, $n^{(k+1)/2}$). Recently, Balogh, Csaba, Jing and Pluhár [1] initiated the study of discrepancies in general graphs. In particular, they obtained a Dirac-type bound for positive discrepancy of Hamilton cycles (in 2 colours; this was generalised to r colours in [7] and independently in [10]), and estimated the discrepancy of the set of all spanning trees in random regular graphs and 2-dimensional grids (in 2 colours). The last result was greatly generalised to r colours and to almost every base graph in [10], where the authors establish a non-trivial connection between the spanning-tree discrepancy (essentially an extremal quantity) and a purely geometric property of the graph. In 2-dimensional grids, Balogh et al. also showed that the discrepancy of paths (hence also of trees) is linear in the number of vertices.

Other recent works include an estimate of multicolour discrepancy in random graphs and in the complete graph [9]; a Dirac-type bound for positive 2-colour discrepancy of k-factors [2]; and a Dirac-type bound for positive 2-colour discrepancy of powers of Hamilton cycles [3]. Finally, Gishboliner, Krivelevich and the second author have introduced a notion of *oriented* discrepancy, and studied the oriented discrepancy of Hamilton cycles in dense and in random graphs [11]. We will elaborate on this matter further.

The present work continues this line of research. Our main result shows that the r-colour discrepancy of the set of all trees in a given tree is linear in the number of leaves of that tree. Let us denote the set of all trees by $\mathcal{T}$. Thus, for a graph G, $\mathcal{D}_r(G, \mathcal{T})$ denotes the r-colour discrepancy of trees in G. For a tree T, denote by $\ell(T)$ the number of leaves in T. As a warm-up example, consider the following two simple cases. Let S_ℓ denote the star with ℓ leaves. It is evident that an optimal colouring is an equipartition of the leaves into the r colour classes, and the most unbalanced tree in this case will be a monochromatic substar. Hence, $\mathcal{D}_r(S_\ell, \mathcal{T}) = (r-1)\left\lceil \frac{\ell}{r} \right\rceil$. Similarly, considering the path P_n on n vertices (so $\ell(P_n) = 2$), an optimal colouring can easily be seen to be any periodic colouring, in which the most unbalanced tree will be a single edge. Hence, $\mathcal{D}_r(P_n, \mathcal{T}) = (r-1)\left\lceil \frac{2}{r} \right\rceil$.

Given the above, a natural guess would be that any tree T satisfies $\mathcal{D}_r(T, \mathcal{T}) = (r-1)\lceil \ell(T)/r \rceil$. It turns out that the above holds for $r = 2$ (see below). For $r \geq 3$, however, this is only (at least asymptotically) a lower bound, and the star demonstrates that it is sharp. Our first and main result gives bounds on $\mathcal{D}_r(T, \mathcal{T})$ in terms of $\ell(T)$.

Theorem 1 (Multicolour discrepancy) *For every $r \geq 2$ and every tree T with ℓ leaves,*

$$\left\lceil (r-1) \cdot \frac{\ell}{r} \right\rceil \leq \mathcal{D}_r(T, \mathcal{T}) \leq \left\lceil (r-1) \cdot \frac{\ell}{2} \right\rceil.$$

In particular, for $r = 2$ we have $\mathcal{D}_2(T, \mathcal{T}) = \lceil \frac{\ell}{2} \rceil$.

We explained earlier why the lower bound in Theorem 1 is sharp (asymptotically and for infinitely many values of ℓ). In Sect. 2, where we prove the theorem, we also prove that the upper bound is sharp (exactly and for every ℓ; see Proposition 1).

Using a classical result of Kleitman and West [12] about the maximum number of leaves in a spanning tree of a graph (sometimes called the **maximum leaf number**), we obtain the following improvement[2] and extension (to any number of colours) of [1, Corollary 7].

Corollary 1 *Let $m, n \geq 2$ be integers and let G be the $m \times n$ grid. Then*

$$\mathcal{D}_r(G, \mathcal{T}) \geq \frac{r-1}{4r} \cdot mn + 1 - 2r.$$

Our next result is in the context of signed/oriented discrepancy. Let us lay a formal ground to state our results. The notion of a **signed hypergraph**, introduced by Shi [15], is an extension of the conventional notion of a hypergraph that allows "negative" vertex-edge incidences. Formally, a signed hypergraph $\mathcal{H}$ is a triple $(\mathcal{V}, \mathcal{E}, \psi)$ where $\mathcal{V}, \mathcal{E}$ are disjoint sets ("vertices" and "hyperedges") and $\psi : \mathcal{V} \times \mathcal{E} \to \{-1, 0, 1\}$ is an **incidence function**. For a hyperedge A we set $f(A) = \sum_{a \in \mathcal{V}} f(a) \cdot \psi(a, A)$, and $|f(A)|$ is called the **imbalance** of A. We define the **signed discrepancy** of $\mathcal{H}$ to be

$$\dot{\mathcal{D}}(\mathcal{H}) = \min_{f:\mathcal{V} \to \{\pm 1\}} \max_{A \in \mathcal{E}} |\sum_{a \in \mathcal{V}} f(a) \cdot \psi(a, A)|.$$

With a slight abuse, we may ignore the formal definition that contains the incidence function, and instead think of sets in a more general way: for each set and each element, the set can contain the element, not contain the element, or "negatively" contain that element. This notion turns out to be useful in many cases, as we will see below. Note that $\dot{\mathcal{D}}(\mathcal{H}) = \mathcal{D}_2(\mathcal{H})$ if ψ is nonnegative; in that sense, the signed discrepancy is a direct generalisation of 2-colour discrepancy.

Analogously to how we defined multicolour discrepancies in graphs, we define oriented discrepancy in graphs. In this setting, given an oriented[3] base graph G and a family of oriented graphs $\mathcal{X}$, we construct a signed hypergraph whose vertices are the edges of G and whose hyperedges are edge sets that from a member of $\mathcal{X}$, where

---

[2] Their result, for $r = 2$ only, is an immediate corollary of a stronger result they prove on the discrepancy of *paths* in the grid. On the other hand, while their proof is a clever ad-hoc and suited for grids, our proof is more general.

[3] That base orientation will not matter and can be arbitrary.

an edge is positively contained in a hyperedge if its orientation in G agrees with its orientation in $\mathcal{X}$, and negatively contained otherwise. The **oriented discrepancy of $\mathcal{X}$ in** G, denoted $\overrightarrow{\mathcal{D}}(G, \mathcal{X})$, is the signed discrepancy of that signed hypergraph.[4] Again, it is convenient to think about this definition in terms of a game: an adversary orients the edges of G (ignoring the "original" orientation it had). He tries to do it as balanced as possible, that is, so that in any member of $\mathcal{X}$, the number of edges in which the orientation in $\mathcal{X}$ agrees with his orientation of G is as close to 50% as possible. Our goal is then to find a member of $\mathcal{X}$ of maximum imbalance, namely, that contains many more agreements than disagreements, or the other way around.

Let $\mathcal{DHAM}$ be the set of all *directed* Hamilton cycles. The result of [11] on the oriented discrepancy of Hamilton cycles in Dirac graphs can be restated as follows: if G is an n-vertex graph with $\delta(G) \geq n/2 + 8$ then $\overrightarrow{\mathcal{D}}(G, \mathcal{DHAM}) = \Omega(2\delta(G) - n))$. The authors of [11] conjectured that if $\delta(G) \geq n/2$ then $\overrightarrow{\mathcal{D}}(G, \mathcal{DHAM}) \geq 2\delta(G) - n$, and that if true, it would be best possible. The conjecture—a strong generalisation of Dirac's theorem—was fully resolved by Freschi and Lo [8].

Here, we obtain a new result in the setting of oriented discrepancy in graphs. Let $\mathcal{DT}$ denote the set of all directed rooted trees; namely, trees that have a distinguished vertex called the **root** and that are oriented *away* from that root.[5] Our next theorem gives bounds on $\overrightarrow{\mathcal{D}}(T, \mathcal{DT})$ in terms of $\ell(T)$.

Theorem 2 (Oriented discrepancy) *For every tree T on at least* 3 *vertices and with ℓ leaves,*

$$\left\lceil \frac{\ell}{2} \right\rceil + 1 \leq \overrightarrow{\mathcal{D}}(T, \mathcal{DT}) \leq \ell.$$

The lower bound is sharp (exactly and for every ℓ), since a star with ℓ leaves that is oriented as evenly as possible has oriented imbalance $\lceil \ell/2 \rceil + 1$. We conjecture that one can obtain a better upper bound that matches the lower bound asymptotically, namely, that $\overrightarrow{\mathcal{D}}(T, \mathcal{DT}) \sim \ell/2$.

The multicolour discrepancy $\mathcal{D}_r$ and the signed discrepancy $\dot{\mathcal{D}}$ are two natural generalisations of the classical notion of discrepancy $\mathcal{D}$, both of combinatorial nature. In some sense, however, they lack the geometric aspect of discrepancy. In particular, $\mathcal{D}_r$ is not even generally monotone in r. Following Tao [16], we may generalise the definition of discrepancy geometrically, by allowing vector-valued colouring functions. Here, we restrict our attention to the (already challenging) case of the vector space $\mathbb{R}^d$. For $d > 0$, let $\mathbb{S}^d$ denote the d-dimensional unit hypersphere in $\mathbb{R}^{d+1}$. A d-dimensional colouring of $\mathcal{H}$ is a function $f : \mathcal{V} \to \mathbb{S}^d$. For a hyperedge A we set $f(A) = \sum_{a \in A} f(a)$, and $|f(A)|$ is called the **imbalance** of A. We define the **d-dimensional discrepancy** of $\mathcal{H}$ to be

[4] A potential term would have been **signed discrepancy**; however, when the vertices of the hypergraph represent edges of a graph, the notion of orientation is more natural.

[5] This is an arbitrary choice of one of two natural orientations of a rooted tree, and has no implications on the results.

$$\mathcal{D}^d(\mathcal{H}) = \min_{f:\mathcal{V}\to\mathbb{S}^d} \max_{A\in\mathcal{E}} |f(A)|.$$

We observe that $\mathcal{D}^0 = \mathcal{D}_2$, and that $\mathcal{D}^{d'} \leq \mathcal{D}^d$ whenever $d' \geq d$. Understanding $\mathcal{D}^0$ quite well, we move on to study $\mathcal{D}^d$ for $d \geq 1$.

While the r-colour discrepancy is more combinatorial in nature and the d-dimensional discrepancy is more geometric, they are related by the following inequality: for every $r \geq 2, d \geq 1$ and hypergraph $\mathcal{H}$, $\mathcal{D}^d(\mathcal{H}) \leq \mathcal{D}^1(\mathcal{H}) \leq \mathcal{D}_r^\circ(\mathcal{H})$. It follows from Eq. 1, Theorem 1 that the high-dimensional tree-discrepancy of a tree with ℓ leaves is at most $\lceil \ell/2 \rceil$ (for every dimension $d \geq 1$). In the next theorem we prove a lower bound that we believe that under some assumptions matches the upper bound. Let $\mathrm{B}(z_1, z_2)$ be the **beta function**.

Theorem 3 (high-dimensional discrepancy) *For every tree T with ℓ leaves,*

$$\mathcal{D}^d(T, \mathcal{T}) \geq \frac{\ell}{d \cdot \mathrm{B}\left(\frac{d}{2}, \frac{1}{2}\right)}.$$

In particular, $\mathcal{D}^1(T, \mathcal{T}) \geq \ell/\pi$ and $\mathcal{D}^d(T, \mathcal{T}) \geq (1 - o_d(1))\ell/\sqrt{2\pi d}$.

We conjecture that $\mathcal{D}^1(T, \mathcal{T}) \geq 1/(2\sin(\pi/(2\ell))) \sim \ell/\pi$ and that when $\Delta(T) \to \infty$, $\mathcal{D}^1(T, \mathcal{T}) \sim \ell/\pi$.

2 Multicolour Discrepancy

In this section we prove Theorem 1 and its sharpness. Given a tree T, a colouring $f : E(T) \to [r]$ and a subtree S, write $e_j(S) = |f^{-1}(S)|$ and $w_j(S) = re_j(S) - |E(S)|$.

Proof (Of the lower bound in Theorem 1) Let T be a tree with ℓ leaves, and let $f : E(T) \to [r]$ be an r-colouring of its edges. Denote $m_j = |f^{-1}(j)|$ for $j \in [r]$ and $m = \sum_{j\in[r]} m_j = |E(T)|$. We obtain a subtree T' of T be deleting all leaves of T, and denote $m'_j = |f^{-1}(j) \cap E(T')|$ and $\ell_j = |f^{-1}(j) \setminus E(T')|$ for $j \in [r]$. Write $m' = \sum_{j\in[r]} m'_j$ and note that $\sum_{j\in[r]} \ell'_j = m - m' = \ell$. Finally, we obtain T_j from T' by adding back the edges $f^{-1}(j) \setminus E(T')$. Observe that $w_j(T_j) = rm_j - m' - \ell_j$, hence

$$\sum_{j\in[r]} w_j(T_j) = rm - rm' - \ell = (r-1)\ell.$$

Thus, by the pigeonhole principle, there exists $j \in [r]$ for which $w_j \geq \lceil (r-1)\ell/r \rceil$, hence $\mathcal{D}_r(T, \mathcal{T}) \geq \lceil (r-1)\ell/r \rceil$.

We move on to prove the upper bound in Theorem 1. Consider the pointwise partial order relation on $\mathbb{R}^r$ defined as follows: for $\mathbf{m} = (m_1, \ldots, m_r)$ and $\mathbf{n} = (n_1, \ldots, n_r)$, $\mathbf{m} \leq \mathbf{n}$ if and only if $m_j \leq n_j$ for every $j \in [r]$. For a permutation τ of $[r]$ we write $\tau(\mathbf{m}) = (m_{\tau(1)}, \ldots, m_{\tau(r)})$. We say that $\mathbf{m}$ is **dominated** by $\mathbf{n}$ and denote it $\mathbf{m} \preceq \mathbf{n}$ if there exists a permutation τ of $[r]$ such that $\tau(\mathbf{m}) \leq \mathbf{n}$. We further write $\mathbf{m} \vee \mathbf{n} = (m_1 \vee n_1, \ldots, m_r \vee n_r)$, where for real numbers $x, y, x \vee y = \max\{x, y\}$. Call $\mathbf{m}$ is **increasing** if it is (weakly) monotone increasing as a sequence. Denote by $\sigma_{\mathbf{m}}$ the first permutation of $[r]$ (according to some arbitrary fixed ordering) for which $\sigma_{\mathbf{m}}(\mathbf{m})$ is increasing. Write $\mathrm{i}(\mathbf{m}) = \sigma_{\mathbf{m}}(\mathbf{m})$ for the "monotone version" of $\mathbf{m}$. Let $\min \mathbf{m}$ and $\max \mathbf{m}$ denote the minimal and maximal coordinate in $\mathbf{m}$, respectively, and note that if $\mathbf{m} \preceq \mathbf{n}$ then $\max \mathbf{m} \leq \max \mathbf{n}$. Say that a vector $\mathbf{m}$ is **1-Lipschitz** if for every $1 \leq j < r$, $|\mathbf{m}_{j+1} - \mathbf{m}_j| \leq 1$.

For a vector $\mathbf{m}$, let $\alpha_{\mathbf{m}}$ be the vector $(a_1, \ldots, a_r)$ where $a_j = m_j + r - \sigma_{\mathbf{m}}^{-1}(j)$. It is useful to observe that if $\mathbf{m} \preceq \mathbf{n}$ then $\alpha_{\mathbf{m}} \preceq \alpha_{\mathbf{n}}$. It is also useful to observe that if $\mathbf{m}$ is increasing 1-Lipschitz then $\alpha_{\mathbf{m}}$ is decreasing 1-Lipschitz. For every $j \in [r]$ denote $d_j = r - \lceil (r + 1 - j)/2 \rceil$, and let $\mathbf{d}_2 = (d_1, \ldots, d_r)$. For $\ell \geq 2$ define $\mathbf{d}_{\ell+1} = \mathrm{i}(\alpha_{\mathbf{d}_\ell})$, and note that for $\ell \geq 3$, $\min \mathbf{d}_\ell = \max \mathbf{d}_{\ell-1}$. Note futher that since $\mathbf{d}_2$ is increasing 1-Lipschitz then by the discussion above, $\mathbf{d}_\ell$ is increasing 1-Lipschitz for every $\ell \geq 2$. Thus, for every $\ell \geq 3$, $(\mathbf{d}_\ell)_j = (\mathbf{d}_{\ell-1})_{r+1-j} + j - 1$. In particular, $\max \mathbf{d}_3 = d_1 + r - 1 = 2r - 1 - \lceil r/2 \rceil = \lceil 3(r-1)/2 \rceil$, and, for $\ell \geq 4$, $\max \mathbf{d}_\ell = \min \mathbf{d}_{\ell-1} + r - 1 = \max \mathbf{d}_{\ell-2} + r - 1$. By induction, $\max \mathbf{d}_\ell = \lceil \ell(r-1)/2 \rceil$. The following claim will be useful for us.

Claim 2.1 *For every $\ell \geq 3$, if $\mathbf{m} \preceq \mathbf{d}_{\ell-1}$ and $\mathbf{n} \preceq \mathbf{d}_\ell$ then $\mathbf{m} \vee \mathbf{n} \preceq \mathbf{d}_\ell$.*

Proof We may assume that $\mathrm{i}(\mathbf{n}) = \mathbf{d}_\ell$. Thus, $\min \mathbf{n} = \min \mathbf{d}_\ell = \max \mathbf{d}_{\ell-1}$, hence $\mathbf{m} \vee \mathbf{n} = \mathbf{n}$, and the claim follows.

Proof (Of the lower bound in Theorem 1) For vertices $u, v \in V(T)$, let $\mathcal{S}_v(T)$ denote the set of subtrees S of T that contain the vertex v, let $\mathcal{S}_{u,v}(T)$ the set of subtrees S that contain u, v, and let $\mathcal{S}_{u,\neg v}(T)$ be the set of subtrees S that contain u but not v. For $j \in [r]$, define $M_j(v; T) = \max_{S \in \mathcal{S}_v(T)} w_j(S)$ and analogously $M_j(u, v; T)$ and $M_j(u, \neg v; T)$. The **colour profile** of v in T (with respect to a colouring f) is the vector $\chi(v; T) = (M_1(v; T), \ldots, M_r(v; T))$. Define analogously $\chi(u, v; T)$ and $\chi(u, \neg v; T)$. We prove by induction the following statement: for every $\ell \geq 2$ and every tree T with ℓ leaves, there exists a colouring f of $E(T)$ for which for every vertex $v \in V(T)$, $\chi(v; T) \preceq \mathbf{d}_\ell$. This would imply, in particular, that for every subtree S of T and every colour $j \in [r]$, $w_j(S) \leq \max \chi(v; T)$ for some vertex $v \in V(S)$; but for every $v \in V(T)$, $\max \chi(v; T) \leq \max \mathbf{d}_\ell \leq \lceil (r-1)\ell/2 \rceil$, implying the statement of the theorem. Our inductive argument yields a concrete explicit colouring of $E(T)$; see Sect. 2 for an (implied) efficient algorithmic version.

The base case is when $\ell = 2$. Here, T is a path; suppose the edges of the path are $(e_1, \ldots, e_k)$ in this order. We colour the path periodically; namely, we let $f(e_i) = j$ if and only if $i \equiv j \pmod r$. Let $v \in V(T)$ and let $S \in \mathcal{S}_v$ be a subpath of T containing v. Evidently, $w_j(S) \leq r - 1$ for every $j \in S$. Thus, $\chi(v; T) \preceq \mathbf{d}_2$.

We move on to the induction step. Let T be a tree with $\ell = \ell(T) \geq 3$ and suppose the statement holds for $\ell - 1$. Let u be a leaf in T, and let b be the **branching vertex** of u, namely, the nearest vertex to u with degree greater than two. Let P_u be the path connecting u to b, and denote by T' the subtree of T obtained by removing all edges of P_u and all vertices of P_u but b. Evidently, $\ell(T') = \ell - 1$. By the induction hypothesis, there exists an r-colouring f' of $E(T')$ that satisfies $\chi(v; T') \preceq \mathbf{d}_{\ell-1}$ for every $v \in V(T')$. We extend f' to a colouring f of $E(T)$ as follows. Let $\mathbf{b}' = \chi(b; T')$ be the colour profile of b in T'. Consider the permutation $\sigma_{\mathbf{b}'}$. Colour the edges of P_u periodically according to $\sigma_{\mathbf{b}'}$; namely, if $P_u = (e_1, \ldots, e_k)$ (where $b \in e_1$ and $u \in e_k$), let $f(e_i) = \sigma_{\mathbf{b}'}(j)$ if and only if $i \equiv j \pmod r$. Note that for any subpath Q of P_u that contains b, and for any colour $j \in [r]$, $w_j(Q) \leq r - \sigma_{\mathbf{b}}^{-1}(j)$. We now show that T satisfies the hypothesis (with respect to f). Namely, we show that for every $v \in V(T)$, $\chi(v; T) \preceq \mathbf{d}_\ell$. We consider three separate cases.

Case I, $v = b$: We observe that for every $S \in \mathcal{S}_b$ and every $j \in [r]$, letting $S' = S \cap T'$ and $S^- = S \cap P_u$, we have $w_j(S) = w_j(S') + w_j(S^-) \leq M_j(b; T') + r - \sigma_{\mathbf{b}'}^{-1}(j)$. Thus, $\chi(b; T) \leq \alpha_{\mathbf{b}'}$. By the induction hypothesis, $\mathbf{b}' \preceq \mathbf{d}_{\ell-1}$, hence $\chi(b; T) \leq \alpha_{\mathbf{b}'} \preceq \mathfrak{i}(\alpha_{\mathbf{d}_{\ell-1}}) = \mathbf{d}_\ell$.

Case II, $v \in V(P_u) \setminus \{b\}$: As with the base case of the induction, we have $\chi(v, \neg b; T) \leq \mathbf{d}_2 \leq \mathbf{d}_{\ell-1}$. On the other hand, $\chi(v, b; T) \leq \chi(b; T) \preceq \mathbf{d}_\ell$ (by Case I). Thus, $\chi(v; T) = \chi(v, \neg b; T) \vee \chi(v, b; T) \preceq \mathbf{d}_\ell$ (by Sect. 2.1).

Case III, $v \in V(T') \setminus \{b\}$: By the induction hypothesis $\chi(v, \neg b; T) \leq \chi(v; T') \preceq \mathbf{d}_{\ell-1}$. On the other hand, $\chi(v, b; T) \leq \chi(b; T) \preceq \mathbf{d}_\ell$. Thus, $\chi(v; T) = \chi(v, \neg b; T) \vee \chi(v, b; T) \preceq \mathbf{d}_\ell$ (by Sect. 2.1).

The proof is now complete.

We briefly discuss how the inductive argument presented in the proof of the upper bound of Theorem 1 yields a simple and efficient algorithm for finding a colouring the achieves at least the upper bound.

We begin by describing an efficient algorithm to compute the colour profile of a vertex b in an r-coloured tree T. The input is a given tree T with m edges, a vertex b, and an r-colouring f. For every vertex v of T, let T_v denote the tree rooted at v comprised of v and all its descendants in T. Now observe that, by considering the imbalance of color j at each subtree T_v, $v \in N(b)$, we have

$$M_j(b; T) = \sum_{v \in N(b)} \max\{r \cdot \mathbf{1}_{f(\{b,v\})=j} - 1 + M_j(v; T_v), 0\}.$$

Hence, computing $\chi(b; T)$ requires $O(rm)$ steps.

We proceed by describing the colouring procedure. We are given a tree T with m edges and ℓ leaves, and a number of colours r. Let u_1, u_2 be two distinct leaves of T, and let P be the unique path between them in T. We colour P alternately with a fixed (arbitrary) cyclic order of the colours. Set $T' = P$. We then iterate over the remaining $\ell - 2$ leaves: given a leaf u that is not in T', let P_u be the unique path in T from u to T', and let b be the last vertex in the path (so $b \in V(T')$). We can calculate the colour profile $\chi(b; T')$ of b in T' in $O(rm)$ steps. Given the colour profile, we colour the path from b to u alternately with a cyclic order of the colours, from the least popular colour up to the most popular. That is, the order of colours is $\sigma_{\chi(b;T')}$. We then add the new coloured path to T' and continue to the next leaf outside T'.

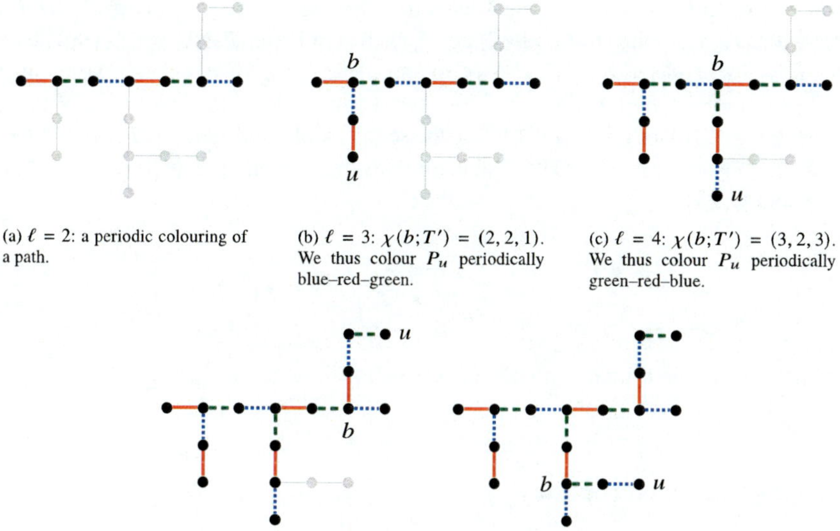

(a) $\ell = 2$: a periodic colouring of a path.

(b) $\ell = 3$: $\chi(b; T') = (2, 2, 1)$. We thus colour P_u periodically blue–red–green.

(c) $\ell = 4$: $\chi(b; T') = (3, 2, 3)$. We thus colour P_u periodically green–red–blue.

(d) $\ell = 5$: $\chi(b; T') = (3, 4, 3)$. We thus colour P_u periodically red–blue–green.

(e) $\ell = 6$: $\chi(b; T') = (5, 3, 4)$. We thus colour P_u periodically green–blue–red.

Fig. 1 Visualisation of the inductive 3-colouring of a tree. The colour profiles are indexed red–green–blue

This algorithm runs, therefore, in $O(rm\ell)$ steps. Its correctness was verified recursively in the proof of the upper bound in Theorem 1. See Fig. 1 for a visualisation of the algorithm.

We now prove Corollary 1.

Proof (Of Corollary 1) Let G be the 2-dimensional $m \times n$ grid ($m, n \geq 2$). Obtain G^+ from G by adding a perfect matching covering the 4 vertices of degree 2 in G, so $\delta(G^+) = 3$. Hence, by [12, Theorem 2], G^+ has a spanning tree T with at least $mn/4 + 2$ leaves. By Theorem 1,

$$\mathcal{D}_r(G^+, \mathcal{T}) \geq \mathcal{D}_r(T, \mathcal{T}) \geq \frac{r-1}{r} \cdot \left(\frac{mn}{4} + 2\right) = \frac{r-1}{4r} \cdot mn + 2 - \frac{2}{r} \geq \frac{r-1}{4r} \cdot mn + 1.$$

The result follows since $\mathcal{D}_r(G, \mathcal{T}) \geq \mathcal{D}_r(G^+, \mathcal{T}) - 2Wr$.

In the introduction, we showed the lower bound of Theorem 1 is asymptotically tight. Here, we show the upper bound is (exactly) tight.

Proposition 1 *For every $r, \ell \geq 2$ there exists a tree T with $\ell(T) = \ell$ and*

$$\mathcal{D}_r(T, \mathcal{T}) = \left\lceil (r-1) \cdot \frac{\ell}{2} \right\rceil.$$

Proof A **spider** is a star-like graph defined as follows: for $k \geq 1$ and $\ell \geq 2$, Sp_ℓ^k is a tree with a root attached to ℓ paths ("legs"), each is of length k. Note that $\ell(\mathrm{Sp}_\ell^k) = \ell$. Let $T = \mathrm{Sp}_\ell^r$, and let $f : E(T) \to [r]$ be an r-colouring of its edges. We identify $E(T)$ by $[\ell] \times [r]$ by labelling the h'th edge (counting from the root) of the i'th leg (i, h). For $j \in [r]$ let $E_j \subseteq [\ell] \times [r]$ be the set of j-coloured edges in T. For $j \in [r]$, let S_j be the smallest subtree of T that contains the root and every j-coloured edge. Note that $w_j(S_j) \geq \sum_{(i,h)\in E_j}(r - h)$, thus

$$\sum_{j\in[r]} w_j(S_j) \geq \sum_{(i,h)\in E(T)} (r - h) = \ell\binom{r}{2}.$$

By the pigeonhole principle, there exists $j \in [r]$ for which $w_j(S_j) \geq \lceil (r - 1)\ell/2 \rceil$.

3 Oriented Discrepancy

In this section we prove Theorem 2 and the sharpness of its lower bound.

Proof (Of the lower bound in Theorem 2) The oriented discrepancy of trees in an ℓ-leaf star is $\lceil \ell/2 \rceil + 1$. We may therefore assume that T is not a star. In particular, there exists a vertex u which is not a leaf and has a neighbour which is also not a leaf. Consider an arbitrary orientation of the edges of T. Consider a 2-colouring of T according to the direction of each edge with respect to u: colour e red if it is oriented towards u, and blue otherwise. By Theorem 1, there exist a subtree T^* of T with 2-colour imbalance at least $\lceil \frac{\ell}{2} \rceil$. Assume T^* maximises the 2-colour imbalance. Assume further, without loss of generality, that the popular colour in T^* is red. In particular, every edge from T^* to its complement is blue. We claim that u is in T^*, and is not a leaf of T^*. Indeed, if u is not in T^*, let w be the closest vertex to u in T^*. Then, since the edge $\{w, z\}$ along the path from w to u is blue, the rooted tree $(T^* + z, w)$ has oriented imbalance at least $\lceil \frac{\ell}{2} \rceil + 1$. Similarly, if u is a leaf of T^*, then, since u has a neighbour v outside T^*, and the edge $\{u, v\}$ is blue, the rooted tree $(T^* + v, v)$ has oriented imbalance at least $\lceil \frac{\ell}{2} \rceil + 1$. This shows, in particular, that all edges incident to u are red.

We conclude that every subtree of T of maximal 2-colour imbalance contains u and its neighbourhood, and that all of these trees have the same popular colour (say, red). By the choice of u, it has a non-leaf neighbour v. Let $e = \{u, v\}$ and consider the tree $T_1 = T/e$ that is obtained from T be contracting e and keeping the original orientations (and the induced 2-colouring). Note that due to the choice of v, u is not a leaf of T_1 and $\ell(T_1) = \ell(T) = \ell$. Thus, applying Theorem 1 again yields a subtree T_1^* of T_1 of maximal 2-colour imbalance, which is at least $\lceil \frac{\ell}{2} \rceil$. By repeating the argument above (in which we did not assume that u has a non-leaf neighbour, but only that it is not a leaf itself), we conclude that u is in T_1^*. Let T_2^* be obtained from T_1^* be de-contracting e. If the dominant colour of T_1^* is blue, then the rooted tree

(T_2^*, v) has oriented discrepancy at least $\lceil \frac{\ell}{2} \rceil + 1$. If the dominant colour of T_1^* is red, then the rooted tree (T_2^*, u) has oriented discrepancy at least $\lceil \frac{\ell}{2} \rceil + 1$.

Proof (Of the upper bound in Theorem 2) For a tree T with a fixed orientation we define $x_{v\rightarrow}^T, x_{v\leftarrow}^T$ to be the largest possible imbalance of a subtree rooted at v in which the dominant orientation is from, respectively to, v. We will prove that there exists an orientation of T such that $x_{v\rightarrow}^T + x_{v\leftarrow}^T \leq \ell$ for all vertices v. We prove the statement by induction on ℓ.

For a tree with two leaves, i.e., a path, orienting the edges alternately clearly works. Assume the statement holds for any tree with up to ℓ leaves and let T be a tree with $\ell + 1$ leaves. Consider a leaf and the path connecting it to its branching vertex v. Let T' be the tree after removing this path. By induction, T' admits an orientation for which $x_{v\rightarrow}^{T'} + x_{v\leftarrow}^{T'} \leq \ell - 1$ for all $v \in T'$. Assume, without loss of generality that $x_{v\rightarrow}^{T'}$ is the smallest of the two. Now, orient the edges of the removed path alternately, where the edge incident to v is oriented towards it. We claim that this orientation of T satisfies the requirements.

First, for any $u \in V(T')$, only $x_{u\leftarrow}^{T'}$ can possibly be increased by 1, so $x_{u\rightarrow}^T + x_{u\leftarrow}^T \leq \ell$. Now for any vertex u in the path, we have

- $(x_{u\rightarrow}^T, x_{u\leftarrow}^T) \leq (x_{v\rightarrow}^{T'} + 2, x_{v\leftarrow}^{T'} - 1)$ if $d(u, v)$ is odd. Note that the only case this is not true is when $x_{v\leftarrow}^{T'} = 0$. But then, $0 \leq x_{v\rightarrow}^{T'} \leq x_{v\leftarrow}^{T'} = 0$, which is impossible, since at least one of them should be positive. Thus, $x_{u\rightarrow}^T + x_{u\leftarrow}^T \leq \ell$.
- $(x_{u\rightarrow}^T, x_{u\leftarrow}^T) \leq (x_{v\rightarrow}^{T'}, x_{v\leftarrow}^{T'} + 1)$ if $d(u, v)$ is even. Thus, $x_{u\rightarrow}^T + x_{u\leftarrow}^T \leq \ell$.

This concludes the induction.

4 High Dimensional Discrepancy

We prove Theorem 3. A proof of the next lemma can be found in [14].

Lemma 1 *Let $d \geq 1$, and let $X = (X_1, \ldots, X_d) \sim \mathsf{Unif}(\mathbb{S}^{d-1})$. Then, the random variable X_1 is distributed on $[-1, 1]$ with density function $f_{X_1}(x) = \frac{(1-x^2)^{\frac{d-3}{2}}}{\mathrm{B}(\frac{d-1}{2}, \frac{1}{2})}$.*

Proof (Of Theorem 3) Let T be a tree with ℓ leaves, and let $f : E(T) \to \mathbb{S}^d$ be a d-dimensional colouring its edges. Let $L \subseteq E(T)$ be the set of edges in T that are incident to a leaf (so $|L| = \ell$). For a vector $\mathbf{v} \in \mathbb{S}^d$ let $L_{\mathbf{v}}$ be the set of edges e in L for which $\mathbf{v} \cdot f(e) > 0$, and let $L'_{\mathbf{v}} = L \setminus L_{\mathbf{v}}$. Denote by $T_{\mathbf{v}}$ the subtree of T with the edge set $E(T) \setminus L'_{\mathbf{v}}$ and by $T'_{\mathbf{v}}$ the subtree of T with the edge set $E(T) \setminus L_{\mathbf{v}}$. For a subtree S of T, write $D(S) = \sum_{e \in E(S)} f(e)$. Write $e_1 = (1, 0, \ldots, 0) \in \mathbb{S}^d$ for the first vector in the standard basis. Let $\mathbf{v}$ be a uniformly random sampled vector in $\mathbb{S}^d$, and set $D_{\mathbf{v}} = \sum_{e \in L} |\mathbf{v} \cdot f(e)|$. By linearity of expectation and by Lemma 1,

$$\mathbb{E}D_{\mathbf{v}} = \ell \cdot \mathbb{E}|\mathbf{v} \cdot e_1| = 2\ell \cdot \int_0^1 \frac{x(1-x^2)^{d/2-1}}{\mathrm{B}\left(\frac{d}{2}, \frac{1}{2}\right)} dx = \frac{2\ell}{d \cdot \mathrm{B}\left(\frac{d}{2}, \frac{1}{2}\right)}.$$

Thus, there exists a vector $\mathbf{v}$ for which $D_{\mathbf{v}} \geq \frac{2\ell}{d \cdot \mathrm{B}\left(\frac{d}{2}, \frac{1}{2}\right)}$. By Cauchy–Schwarz we get

$$\left\| D(T_{\mathbf{v}}) - D(T'_{\mathbf{v}}) \right\| \geq \left| \mathbf{v} \cdot \left(D(T_{\mathbf{v}}) - D(T'_{\mathbf{v}}) \right) \right| = D_{\mathbf{v}} \geq \frac{2\ell}{d \cdot \mathrm{B}\left(\frac{d}{2}, \frac{1}{2}\right)}.$$

By the triangle inequality, $D(S) \geq \ell / \left(d \cdot \mathrm{B}\left(\frac{d}{2}, \frac{1}{2}\right)\right)$ for some $S \in \{T, T'\}$. A straightforward application of Stirling's formula yields the asymptotic bound as $d \to \infty$.

Acknowledgements The second author wishes to thank Boris Bukh, Matan Harel and Yinon Spinka for fruitful discussions at various stages of this project.

References

1. Balogh, J., Csaba, B., Jing, Y., Pluhár, A.: On the discrepancies of graphs. Electr. J. Combin. **27**, 2, Paper No. 2.12, 14 (2020). https://doi.org/10.37236/8425 MR4245067
2. Balogh, J., Csaba, B., Pluhár, A., Treglown, A.: A discrepancy version of the Hajnal-Szemerédi theorem, 2021, 0963–5483. Combin. Probab. Comput. **30**, 3, 444–459 (2021). https://doi.org/10.1017/s0963548320000516, MR4247634
3. Bradač, D.: Powers of Hamilton cycles of high discrepancy are unavoidable. Electr. J. Combin. **29**, 3, Paper No. 3.22, 26 (2022). https://doi-org.cmu.idm.oclc.org/10.37236/10279MR4458145
4. Doerr, B., Srivastav, A.: Multicolour discrepancies, 2003, 0963-5483. Combin. Probab. Comput. **12**, 4, 365–399 (2003). https://doi.org/10.1017/S0963548303005662, MR1994100
5. Erdős, P., Füredi, Z., Loebl, M., Sós, V.T.: Discrepancy of trees, 1995, 0081-6906. Studia Scientiarum Mathematicarum Hungarica **30**, 1–2, 47–57, MR1341566 (1995)
6. Erdős, P., Spencer, J.H.: Imbalances in k-colorations, 1971/72, 0028-3045. Networks **1**, 379–385 (1971/92). https://doi.org/10.1002/net.3230010407, MR299525
7. Freschi, A., Hyde, J., Lada, J., Treglown, A.: A note on color-bias Hamilton cycles in dense graphs, 2021, 0895-4801. SIAM J. Disc. Math. **35**, 2, 970–975 (2021). https://doi-org.cmu.idm.oclc.org/10.1137/20M1378983, MR4256086
8. Freschi, A., Lo, A.: An oriented discrepancy version of Dirac's theorem, 2022-11, arXiv e-prints, eprint=2211.06950
9. Gishboliner, L., Krivelevich, M., Michaeli, P.: Color-biased Hamilton cycles in random graphs, 2022, 1042-9832. Random Struct. Algorithms **60**, 3, 289–307 (2022). https://doi-org.cmu.idm.oclc.org/10.1002/rsa.21043, MR4388697
10. Gishboliner, L., Krivelevich, M., Michaeli, P.: Discrepancies of spanning trees and Hamilton cycles, 2022, 0095-8956. J. Comb. Theory Ser B **154**, 262–291 (2022). https://doi-org.cmu.idm.oclc.org/10.1016/j.jctb.2022.01.003, MR4374842
11. Gishboliner, L., Krivelevich, M., Michaeli, P.: Oriented discrepancy of Hamilton cycles, 2023, 0364-9024,1097-0118. J. Graph Theory **103**, 4, 780–792 (2023). https://doi.org/10.1002/jgt.22947, MR4606422
12. Kleitman, D.J., West, D.B.: Spanning trees with many leaves, 1991, 0895-4801. SIAM J. Disc. Math. **4**, 1, 99–106 (1991). https://doi-org.cmu.idm.oclc.org/10.1137/0404010, MR1090293
13. Matoušek, J.: Geometric discrepancy. Algorithms Combinatorics, vol. 18, 3-540-65528-X. Springer, Berlin (1999). https://doi.org/10.1007/978-3-642-03942-3, MR1697825
14. Pavlyk, O.: Random point uniform on a sphere (answer) (2012). https://math.stackexchange.com/q/185312

15. Shi, C.-J.: A signed hypergraph model of the constrained via minimization problem. Microelectron. J. **23**, 7, 533–542 (1992)
16. Tao, T.: The Erdős discrepancy problem. Discrete Analysis, Paper No. 1, 29 (2016). https://doi-org.cmu.idm.oclc.org/10.19086/da.609, MR3533300

Handling Sub-symmetry in Integer Programming using Activation Handlers

Christopher Hojny, Tom Verhoeff, and Sten Wessel

Abstract Symmetry in integer programs (IPs) can be exploited to reduce solving times. Usually only symmetries of the original IP are handled, but new symmetries may arise at some nodes of the branch-and-bound tree. While symmetry-handling inequalities (SHIs) can easily be used to handle original symmetries, handling sub-symmetries arising later on is more intricate. To handle sub-symmetries, it has been proposed to add SHIs that are activated by auxiliary variables. But this may increase the IP's size substantially as all sub-symmetries need to be modeled explicitly. We propose an alternative framework for generically activating SHIs, so-called *activation handlers*. In this framework, we define a callback that checks for active sub-symmetries, eliminating the need for auxiliary variables. In particular, activation handlers can activate symmetry-handling techniques that are more powerful than SHIs. We show that our approach is flexible, with applications in the multiple-knapsack and unit commitment problems. Numerical results show a substantial performance improvement on the existing sub-symmetry-handling methods.

1 Introduction

Branch-and-bound (B&B) is a popular method to solve integer programs (IPs). By iteratively splitting IPs into smaller subproblems, B&B can solve problems with thousands of variables and constraints in adequate time, but it is well-known that the presence of symmetries leads to unnecessarily large B&B trees. The main reason is that B&B explores symmetric subproblems, which all provide essentially the same

C. Hojny · T. Verhoeff · S. Wessel (✉)
Eindhoven University of Technology, Eindhoven, The Netherlands
e-mail: s.wessel@tue.nl

C. Hojny
e-mail: c.hojny@tue.nl

T. Verhoeff
e-mail: t.verhoeff@tue.nl

A. Brieden et al. (eds.), *Graphs and Combinatorial Optimization: from Theory to Applications*, AIRO Springer Series 13,
https://doi.org/10.1007/978-3-031-46826-1_8

information. Therefore, symmetry-handling is an important ingredient of modern B&B implementations that substantially improves the running time [23].

We consider permutation symmetries of binary programs $\max\{c^\top x \mid x \in \mathcal{X}\}$, where $c \in \mathbb{Z}^d$ and $\mathcal{X} \subseteq \{0, 1\}^d$. A *permutation* is a bijection of $[d] := \{1, \ldots, d\}$; the set of all permutations is Sym_d. We assume that a permutation $\pi \in \mathrm{Sym}_d$ acts on $x \in \{0, 1\}^d$ by permuting its coordinates, i.e., $\pi(x) := (x_{\pi^{-1}(1)}, \ldots, x_{\pi^{-1}(d)})$. A *symmetry* of the binary program is a permutation $\pi \in \mathrm{Sym}_d$ that preserves the objective, i.e., $c^\top \pi(x) = c^\top x$, and feasibility, i.e., $x \in \mathcal{X}$ if and only if $\pi(x) \in \mathcal{X}$. Note that the set of all symmetries forms a group under composition. We refer to this group as the *symmetry group* of the binary program, denoted by G. Since computing G is NP-hard [19], one usually only handles a subgroup of G, which can either be detected automatically [23, 25] or is provided by an expert.

The *orbit* of x is $\mathrm{orb}_G(x) = \{\pi(x) \mid \pi \in G\}$ and contains all solutions equivalent to x w.r.t. G. Note that orbits partition $\mathcal{X}$. To handle symmetries, it suffices to restrict $\mathcal{X}$ to (usually lexicographically maximal) representatives of orbits. A vector $y \in \mathbb{Z}^d$ is *lexicographically greater than* $z \in \mathbb{Z}^d$, denoted $y \succ z$, if there is $i \in [d]$ with $y_i > z_i$ and $y_j = z_j$ for all $j < i$. We write $y \succeq z$ when $y \succ z$ or $y = z$ hold. A solution $x \in \mathcal{X}$ is *lexicographically maximal* in its orbit under G if $x \succeq \pi(x)$ for all $\pi \in G$.

To solve binary programs, B&B generates subproblems $\max\{c^\top x \mid x \in Q\}$, where $Q \subseteq \mathcal{X}$. Each subproblem is a binary program with symmetry group G_Q. In general, G_Q is different from G and neither is a subgroup of the other. Following [2], we call symmetries in G_Q *sub-symmetries* of the initial binary program. If sub-symmetries appear frequently during the solving process, it can be beneficial to handle them. But since computing (subgroups of) G_Q might be costly, providing knowledge about sub-symmetries via experts can substantially reduce the complexity of handling sub-symmetries. This, however, leads to a new challenge: how to efficiently provide expert knowledge to a binary programming solver.

Recently, [2] suggested to introduce, for each possible sub-symmetry, a simple symmetry-handling inequality (SHI) that is coupled with auxiliary variables that enable (resp. disable) the SHI if the sub-symmetry is active (resp. inactive) at a subproblem. This, however, might lead to a significant increase in the size of the problem formulation as all sub-symmetries need to be explicitly modeled. Moreover, simple SHIs might not lead to the strongest symmetry reductions.

In this paper, we propose an alternative approach. Instead of using auxiliary variables to activate SHIs, we aim to both (i) detect sub-symmetries valid at a B&B node and (ii) handle these symmetries on-the-fly. Our approach is inspired by the celebrated idea to separate (exponentially) large families of cutting planes for IPs instead of adding all of them initially. Such separation routines are usually implemented via callback mechanisms of IP solvers such as SCIP, Gurobi, CPLEX, or Xpress. Therefore, we introduce a new type of callback, so-called *activation handlers*, which is able to detect sub-symmetries and activate sophisticated symmetry handling methods. The aim of this article is to answer the following research question:

Does the success of using callbacks for separating cutting planes carry over to symmetry handling? In particular, to which extent can we improve upon the SHI approach?

In the remainder of this section, we provide a brief overview of symmetry handling methods. Sect. 2 summarizes the state-of-the-art of handling sub-symmetries, which is complemented by a description of our activation handler framework in Sect. 3. Then, we illustrate for three classes of problems how activation handlers can be used to handle sub-symmetries (Sect. 4), and we evaluate our activation handler framework on a broad set of instances (Sect. 5). Numerical results show that our novel framework substantially improves upon the state-of-the-art in handling sub-symmetries.

Related Literature

Many symmetry-handling methods exist in the literature, including variable branching and fixing rules [18, 22], pruning rules [17, 18, 21], model reformulation techniques [5], and symmetry-handling constraints [6, 9, 10, 12–16, 26]. In the following, we discuss some constraint-based techniques that we also use in our experiments. For details on other techniques, we refer to the surveys [19, 23].

Most symmetry-handling constraints enforce that only lexicographically maximal representatives of symmetric solutions are computed. For a single symmetry $\pi \in G$, $x \succeq \pi(x)$ can be enforced in linear time by propagation and separation techniques [4, 10]. For general groups, however, it is coNP-complete to decide whether a solution is lexicographically maximal in its orbit if G is given by a set of generators, c.f. [1]. Attention has thus been spent on groups that arise frequently in practice. One such case assumes that the variables are organized in a matrix $x = (x_{i,j})_{i \in [m], j \in [n]}$ and that the symmetries in G permute the columns of x arbitrarily. Such symmetries arise frequently in benchmark instances [23], and in Sect. 4, we illustrate some applications. There, we also discuss how orbitopes can be used to handle these symmetries.

The *full orbitope* is the convex hull of all binary matrices with lexicographically non-increasingly sorted columns. If restricting to matrices all of whose rows have at most (resp. exactly) one 1-entry, the corresponding convex hull is called *packing* (resp. *partitioning*) *orbitope*. These matrix symmetries can be handled by separating valid inequalities for orbitopes. For packing/partitioning orbitopes, a facet description can be separated in linear time [12]; for full orbitopes, efficiently separable IP formulations are known [10]. Moreover, efficient propagation algorithms for full and packing/partitioning orbitopes are known [3, 11]. The so-called *orbitopal fixing* algorithms receive local variable bounds at a subproblem and derive variables that need to be fixed at a certain value to guarantee that a solution is within the orbitope.

2 Sub-symmetry in Integer Programming

In this section, we provide a detailed explanation of sub-symmetries and intricacies when dealing with them. We use the multiple knapsack problem as a running example for illustration purposes.

The *multiple knapsack problem* (MKP) considers m items with associated profit p_i and weight w_i, $i \in [m]$, as well as n knapsacks with capacity c_j, $j \in [n]$. The objective is to assign each item to at most one knapsack such that the total weight of items assigned to knapsack $j \in [n]$ does not exceed c_j and the total profit of assigned items is maximized. The MKP is NP-hard as it generalizes the NP-hard single-knapsack problem [8]. A standard IP formulation (1) is given below. There, variable $y_{i,j}$ indicates whether item $i \in [m]$ is packed in knapsack $j \in [n]$, see [20]. Let $\mathcal{Y}_{\text{MKP}}$ denote the set of all feasible solution matrices y.

This formulation exhibits two types of symmetries. For any feasible solution, permuting indices of knapsacks with equal capacity yields another feasible solution. When all knapsacks have the same capacity, this symmetry corresponds to permuting the columns of the solution matrix y. Symmetries also arise from items with identical properties (same weight and profit). In any feasible solution such items can be permuted, corresponding to permuting the respective rows in the solution matrix y.

Formulation (1) also admits sub-symmetries: Consider two knapsacks j and j', and an item with index i. Suppose items $\{1, \dots, i-1\}$ are placed such that the remaining capacity of knapsacks j and j' are equal. Then, the placement of the remaining items $\{i, \dots, m\}$ can be permuted between the two knapsacks. We call this type of sub-symmetry *capacity sub-symmetries*. By this definition, also knapsacks j and j' with $c_j \neq c_{j'}$ can become sub-symmetric.

$$\max_{y \in \{0,1\}^{m \times n}} \left\{ \sum_{i=1}^{m} \sum_{j=1}^{n} p_i y_{i,j} \;\middle|\; w^\top y \leq c^\top \text{and} \sum_{j=1}^{n} y_{i,j} \leq 1 \text{ for all } i \in [m] \right\} \tag{1}$$

In general, sub-symmetries can be defined for arbitrary collections of subproblems $\mathbb{S} = \{Q_s \subset \mathcal{X} | s \in [q]\}$ for some q. The sub-symmetries then correspond to permutations in G_{Q_s}, $s \in [q]$, and are either automatically detected by an IP solver or are provided by a user. In the MKP, the solution subsets for which capacity sub-symmetries occur, are defined as

$$Q^i_{j,j'} = \left\{ y \in \mathcal{Y}_{\text{MKP}} \;\middle|\; c_j - \sum_{k=1}^{i-1} w_k y_{k,j} = c_{j'} - \sum_{k=1}^{i-1} w_k y_{k,j'} \right\}, \tag{2}$$

for all pairs $j, j' \in [n]$, $j < j'$, and all $i \in [m]$. We denote by $\mathbb{S}_{\text{MKP}}$ the collection of all these solution subsets.

In B&B, a subproblem (node of the B&B tree) may belong to one or multiple solution subsets Q_s. We then say that the sub-symmetries in G_{Q_s} become *active*. To exploit sub-symmetries, a solver needs to detect when it is in a subproblem where the sub-symmetry is active. Then, the sub-symmetry must be handled, which can be done using methods similar to those for handling global symmetries.

As observed by [3], sub-symmetry handling for different sub-problems might be conflicting. Consider, for example, the MKP for two knapsacks with capacities $c_1 = 6$ and $c_2 = 2$, and three items with weights $(4, 2, 2)$ and profits $(2, 3, 3)$. Initially, only items 2 and 3 are equivalent. This symmetry can be handled by $(y_{2,1}, y_{3,1}, y_{2,2}, y_{3,2}) \succeq (y_{3,1}, y_{2,1}, y_{3,2}, y_{2,2})$, i.e., we prefer item 2 in knapsack 1. If we branch on $y_{1,1} = 1$, both knapsacks have the same remaining capacity. The knapsack sub-symmetry can be handled by $(y_{3,1}, y_{2,1}, y_{2,2}, y_{3,2}) \succeq (y_{3,2}, y_{2,2}, y_{3,1}, y_{2,1})$, i.e, we prefer using knapsack 1 and item 3 first. These constraints, however, allow to pack at most one of the remaining items, which is suboptimal. Therefore, [3] conclude that compatible sub-symmetry handling methods need to be selected.

If one only handles symmetries arising from permutations of columns in the matrix of binary variables, the following structure in the set of sub-symmetries $\mathbb{S}$ ensures compatability [3]. Let $Q \in \mathbb{S}$. For a solution matrix $x \in Q$, let $x(R, C)$ denote the submatrix of x obtained by restricting to rows $R \subseteq [m]$ and columns $C \subseteq [n]$. The symmetry group G_Q is the *sub-symmetric group* with respect to (R, C) if it contains all the permutations of the columns of $x(R, C)$. If G_Q is the sub-symmetric group, then Q is called *sub-symmetric* with respect to (R, C). Now, let $\mathbb{S}$ be a set of solution subsets such that every $Q_s \in \mathbb{S}$ is sub-symmetric with respect to (R_s, C_s). For every orbit σ_s^i of G_{Q_s}, choose the representative $x_s^i \in \sigma_s^i$ such that the submatrix $x_s^i(R_s, C_S)$ is lexicographically maximal in its orbit, i.e., its columns are lexicographically non-increasing. Then, these representatives are compatible [3].

For the MKP example, the capacity sub-symmetries arise in the solution subsets $\mathbb{S}_{\text{MKP}}$. A solution subset $Q^i_{j,j'} \in \mathbb{S}_{\text{MKP}}$ is sub-symmetric with respect to $(\{i, \dots, m\}, \{j, j'\})$. Whenever the sub-symmetry is active, one can thus handle it by enforcing that the columns of the submatrix $y(\{i, \dots, m\}, \{j, j'\})$ are lexicographically non-increasing.

The state-of-the-art approach to handle a sub-symmetry is to add sub-symmetry-handling inequalities to the model [2]. For the MKP, the inequalities are of the form $y_{i,j'} \leq z + y_{i,j}$ for the sub-symmetry in $Q^i_{j,j'}$. When $z = 0$, the inequality partially breaks the symmetry by ensuring correct lexicographical ordering of two entries in the matrix of variables. When $z \geq 1$, the inequality is trivially satisfied. Here, z is an auxiliary non-negative integer variable that is necessary to only activate the SHI whenever we are indeed in the subproblem $Q^i_{j,j'}$, and can be defined as $z = \left|c_j - c_{j'} - \sum_{k=1}^{i-1} w_k(y_{k,j} - y_{k,j'})\right|$. To include this in the IP, we use standard techniques to linearize the expression, leading to additional variables and constraints. We refer to [2] for more details on SHIs and the techniques to make the set of SHIs fully break the sub-symmetry.

3 Activation Handler

The existing method of handling sub-symmetries with inequalities has a number of limitations. For every sub-symmetry that we want to handle, it is necessary to add many explicit SHIs to the formulation, leading to a blow-up of the IP. The size of the formulation increases even more for problems where additional variables or constraints are necessary to express the auxiliary z-variable in the formulation. Additionally, the inequalities are rather weak for symmetry handling. In particular, the variable-based approach is not immediately able to activate more sophisticated symmetry-handling methods such as orbitopal fixing.

To circumvent these issues, we introduce a new approach for handling sub-symmetries. Our framework is defined by the following steps.

S1 Identify the sub-symmetries present in the formulation.
S2 Define how active sub-symmetries can be detected at the nodes of the B&B tree.
S3 Implement a callback for detecting sub-symmetries.
S4 Use a symmetry-handling method to handle the detected sub-symmetries.

S1, S2, and S3 require expert knowledge of the modeled problem, to provide symmetry information of the problem to the callback. S4 allows the modeler to specify a method to handle detected sub-symmetries from within the callback.

This approach makes the activation and handling of sub-symmetries flexible. In the activation handler, complete information about the current node in the B&B tree is available to the modeler. In many cases, such as for the MKP and the problems in Sect. 4, active sub-symmetries can be detected by inspecting variable fixings at a node. Furthermore, the activation handler can be used with any symmetry-handling method from the literature, and is not restricted to inequality-based approaches. Neither activation nor symmetry handling needs to be encoded in the formulation directly, keeping the IP compact.

The validity of the approach depends firstly on the validity of the identified sub-symmetries in S1, for which the arguments are problem-specific. For S4, existing symmetry-handling techniques are used, for which the correctness proofs remain valid in our case. By selecting compatible representatives, handling multiple sub-symmetries simultaneously does not lead to conflicts, which is proven in [3].

As the framework is rather generic, we will illustrate the concept for two problems in Sect. 4. Afterwards, we compare the numerical performance in Sect. 5. Our implementation of the activation handler framework and experiments are publicly available [28]. The generic framework can easily be adapted by practitioners for use in other applications.

4 Application

In this section, we discuss how sub-symmetries arise for two types of problems: the multiple knapsack problem and the unit commitment problem, see also [29] for further applications. We describe how SHIs can be applied to handle sub-symmetries, as well as the activation handler framework. The activation handler uses information from the solver about variable fixings at a node of the B&B tree. To this end, we define for a node a of the B&B tree the sets F_0^a and F_1^a, which denote the variables that are fixed to 0 or 1 at node a, respectively.

4.1 Multiple Knapsack Problem

We introduced the MKP, its symmetries, and the capacity sub-symmetries in Sect. 2. Notice that the set of solution subsets $\mathbb{S}_{\mathrm{MKP}}$, which considers all pairs of knapsacks, can be generalized to arbitrary groups of knapsacks with identical residual capacities. When handling the symmetry with inequalities, considering all these solution subsets is intractable, as potentially every subset of knapsacks might define a sub-symmetry. That is, exponentially many SHIs as well as auxiliary variables and constraints need to be added to the problem. Therefore, we only consider SHIs for *consecutive* pairs of knapsacks, i.e., $j' = j + 1$. We hence add $\mathcal{O}(mn)$ SHIs to the formulation, with for each SHI four auxiliary variables and five auxiliary constraints.

Sub-symmetry-handling with Activation Handler

When handling sub-symmetries via activation handlers, we are more flexible in implementing the activation rules. Instead of enumerating every solution subset $Q^i_{j_1,j_2}$ separately and checking if the sub-symmetry is active, we can use a single activation handler in the model. The activation handler returns all submatrices of y that contain active sub-symmetries at a given node of the B&B tree.

The activation handler identifies whether the placement of items $[i-1]$ is fixed at node a, according to the variable fixings F^a_0 and F^a_1. For every item i for which the previous holds, the activation handler checks whether there are knapsacks of equal remaining capacity, after placement of items $[i-1]$. Suppose that for item i the knapsacks $j_{k_1}, \dots, j_{k_r}$ have equal remaining capacity. Then, the activation handler reports the submatrix $y(\{i, \dots, m\}, \{j_{k_1}, \dots, j_{k_r}\})$, for which the capacity sub-symmetry is now active at node a. In this way, finding all active capacity sub-symmetries is linear in the number of variables in the matrix y, as we can simply perform a linear scan over the rows of y and checking the variable fixings. Note that checking whether a variable is fixed can be done in constant time, as this information is available from the solver, through our new callback interface.

The activated submatrices are then passed to a high-level symmetry-handling constraint in the solver. Several methods can be used to handle symmetry in the submatrix. In our implementation, we use orbitopal fixing for packing orbitopes [11] to handle the active sub-symmetries.

4.2 Unit Commitment Problem

Another problem in which we can handle sub-symmetries is the min-up/min-down unit commitment problem (MUCP), as introduced in [2]. We are given a set of production units $\mathcal{U}$ with $|\mathcal{U}| = n$, and a discrete time horizon $\mathcal{T} = [T]$ for which a certain non-negative demand D_t needs to be satisfied at every time $t \in \mathcal{T}$. Every production unit $j \in \mathcal{U}$ can be either *up* or *down* at every $t \in \mathcal{T}$. When a unit is up, its production is between a minimum and maximum production capacity $P^j_{\min}$ and $P^j_{\max}$, and it must remain up for at least L^j time steps. When a unit is down, its production is zero and it must remain down for at least ℓ^j time steps. We furthermore have for every unit j a start-up cost c_0^j, a fixed cost c_f^j for every time step the unit is up, and a production cost c_p^j proportional to its production. The goal is to find a *production schedule* satisfying the production demand at every time step and the min-up and min-down constraints, while minimizing the total cost.

Let the variables $x_{t,j} \in \{0, 1\}$ indicate whether unit $j \in \mathcal{U}$ is up at time $t \in \mathcal{T}$, and $u_{t,j} \in \{0, 1\}$ whether unit j starts up at time t. We omit further details of the IP formulation we use for this problem, as they are not relevant for symmetry handling, and refer to [2] for details. Let $\mathcal{X}_{\text{MUCP}}$ denote the set of matrices $(x_{t,j})$ that are feasible. Notice that the solution matrix x completely characterizes a solution, as the corresponding matrix u can be derived completely from x.

Symmetries are present globally in the MUCP when production units have identical properties, i.e., units where the properties $(P_{\min}, P_{\max}, L, \ell, c_0, c_f, c_p)$ are equal. For now, we assume that all units are indeed identical, and we can hence permute their production schedules. This corresponds to permuting the columns of x. One possible way of breaking the symmetry is to restrict x to the full orbitope for binary matrices of size $T \times n$, i.e., by imposing that the columns of x are lexicographically non-increasing. If not all units are identical, this approach can instead be applied to submatrices x that only contain the columns corresponding to identical units.

The MUCP also exhibits sub-symmetries, as introduced in [2]. Call a production unit $j \in \mathcal{U}$ *ready to start up* at some time $t \in \mathcal{T}$ if the unit has been down continuously for at least the minimum downtime ℓ^j. In other words, when $x_{t',j} = 0$ for all $t' = t - \ell^j, \ldots, t - 1$ and $t \geq \ell^j + 1$. Now, suppose there are at least two units $j_1, \ldots, j_k \in \mathcal{U}$ that are all ready to start up at some time t. Then, their production schedules can be permuted from time t onwards, regardless of their schedule up to time t. This thus defines a sub-symmetry where the columns of the submatrix $x(\{t, \ldots, T\}, \{j_1, \ldots, j_k\})$ can be permuted. Analogously, one can identify sub-symmetries for units ready to shut down at some time $t \in \mathcal{T}$. We refer to these sub-symmetries as *start-up* and *shut-down sub-symmetries*, respectively.

Sub-symmetry-handling Inequalities

Following the approach in [2], the start-up sub-symmetries can be handled with inequalities as follows. The handling of shut-down sub-symmetries is analogous, and we omit the details here. Let $j := j_k$, $j' := j_{k+1}$ be a pair of consecutive units, for the sake of brevity. Then, the solution subsets

$$\check{Q}_k^t = \left\{x \in \mathcal{X}_{\text{MUCP}} \,\middle|\, x_{t',j} = x_{t',j'} = 0 \text{ for all } t' = t - \ell, \ldots, t - 1\right\} \tag{3}$$

for all $t \geq \ell + 1$, define when the start-up sub-symmetries occur. For $\check{Q}_k^t$, the corresponding auxiliary variable can be expressed as $z = \sum_{t'=t-\ell}^{t-1} [x_{t',j} + x_{t',j'}]$, leading to the SHI $x_{t,j'} \leq z + x_{t,j}$. Note that the z-variable has a linear description in x, and hence it is not necessary to add z explicitly to the formulation. Instead, we can simply replace z directly with its linear expression. The SHIs for the start-up sub-symmetries can be slightly strengthened, for which we refer to [2] for details on the derivation. A similar inequality can be obtained for the shut-down sub-symmetries.

Sub-symmetry-handling with Activation Handler

Handling sub-symmetry with an activation handler is similar to the approach for the MKP. We add a single activation handler to the model, that identifies all submatrices corresponding to active sub-symmetries in the following manner. For sake of presentation, we assume that all production units $\mathcal{U}$ have the same *type*, i.e., all units have identical properties. In the more general case where we have multiple types of production units, we can simply apply our method to the unit types separately.

Let a be a node of the B&B tree. Define for every $t \in \{\ell + 1, \dots, T\}$,

$$\check{S}_t^a = \left\{ j \in \mathcal{U} \,\middle|\, x_{t',j} \in F_0^a \text{ for all } t' \in \{t - \ell, \dots, t - 1\} \right\}. \tag{4}$$

That is, $\check{S}_t^a$ are the production units that are *fixed* to be ready to start up at time t at node a. For every subset $\check{S}_t^a$ for which $|\check{S}_t^a| \geq 2$, the corresponding start-up sub-symmetry becomes active. Hence, the symmetry corresponds to column permutations of the submatrix $x(\{t, \dots, T\}, \check{S}_t^a)$. We then use orbitopal fixing for full orbitopes to handle the sub-symmetry in the activated submatrix.

Notice that we can find the units that are ready to start up for every time $t \in \mathcal{T}$ in $\mathcal{O}(nT)$ time, by iterating over the time horizon and a dynamic-programming approach. The shut-down sub-symmetries are activated and handled with an analogous approach.

5 Experimental Results

In this section, we compare the sub-symmetry-handling methods using experiments on instances of the problems introduced above.

Instances

Let $U(\ell; L)$ denote a random variable uniformly distributed on $\{\ell, \dots, L\}$. For the MKP, we generate random instances from four standard problem classes [7, 20, 24]:

- *uncorrelated*, $w_i, p_i \in U(\ell; L)$,
- *weakly correlated*, $w_i \in U(\ell; L)$ and $p_i \in U(\max\{1, w_i - \frac{L-\ell}{10}\}; w_i + {}^{L-\ell}/_{10})$,
- *strongly correlated*, $w_i \in U(\ell; L)$ and $p_i = w_i + (L - \ell)/10$,
- *multiple subset-sum*, $w_i \in U(\ell; L)$ and $p_i = w_i$.

We generate our instances with $\ell = 10, L = 1000$. The capacity of every knapsack is set to $c_j = \left\lfloor \frac{1}{2} \sum_{i=1}^{m} w_i / n \right\rfloor$, i.e., the total capacity is approximately half of the total weight of all items. To introduce symmetry in the problem, we generate multiple items with the same weight with an approach similar to [2]. Generated weights are duplicated d times, where $d \in U(1; fm)$, where we call $f \in \{\frac{1}{2}, \frac{1}{3}, \frac{1}{4}, \frac{1}{8}\}$ the *symmetry factor*. Larger values of f generate larger groups of items with equal weight, leading to a more symmetric instance. For generating the profit values for the items within an equal-weight group, we consider two types of instances:

- *equal profit*, where every item in the equal-weight group also has equal profit, generated according to the item class above,
- *free profit*, where every item in the equal-weight group has a profit value generated according to the item class above.

For *strongly correlated* and *multiple subset-sum*, we only generate instances for *equal profit* as both types are equivalent. We make groups of equal-weight items until we have generated m items, $(m, n) \in \{(48, 12), (60, 10), (60, 30), (75, 15), (100, 10)\}$. We generate 20 instances for every combination of item class, symmetry factor, and duplication type, yielding 2400 instances. For MUCP, we use the instances from [2].

Experimental Setup

All experiments are run with a development version of SCIP 7.0.3 (Git hash `3671128c`) with the SoPlex 6.0.0 LP solver [27], on a single core of an Intel Xeon Platinum 8260 CPU (2.4 GHz), with a memory limit of 10 GB of RAM and a time limit of 3600 s. The IP model is constructed in Python 3.10 using the PySCIPOpt interface that exposes the SCIP API in Python. Activation handlers are implemented in SCIP as a new plugin, and can be added via the PySCIPOpt interface.

In order to compare performance of the different symmetry-handling methods, we use the following settings:

- No-Sym: Formulation with SCIP internal symmetry handling turned off.
- Default: Formulation with SCIP default parameters.
- Orbitope: Formulation with orbitope constraints for (global) symmetry handling.
- Ineq: Formulation with SHIs.
- Act: Formulation with orbitope constraints for (global) symmetry handling and activation handler for sub-symmetries.

For the MKP, all models except for No-Sym include orbitope constraints for handling symmetry between identical items. In the orbitopes for symmetries between identical items and symmetries between identical knapsacks, we use a compatible ordering of the variables such that orbitopal fixing for all orbitopes can be performed simultaneously, without introducing any conflicts. For the MUCP, we use the strengthened SHIs in the Ineq model [2]. The orbitope constraints in SCIP use orbitopal fixing, as discussed in Sect. 1.

Results

The results are summarized in Table 1. We aggregate the results for instances in classes, based on the solving time of the tested models. Notation $[a, b)$ denotes the set of instances for which all models have a solving time of at least a and below b seconds. We exclude 551 MKP and 12 MUCP instances from our test set, where all models reach the time limit. For each instance class, we report the number of instances in the class (#). For every model, the number of instances solved to optimality (Opt) and mean solving time (in seconds) is reported. The mean solving time is the shifted

Table 1 Summarized numerical results for MKP and MUCP

Problem	Instances	#	*Existing methods*								*Our method*	
			No-Sym		Default		Ineq		Orbitope		Act	
			Opt	Time	Opt	Time	Opt	Time	Opt	Time	Opt	Time
MKP	All	1849	1166	50.4	1387	33.6	625	1109.6	1636	16.6	1828	12.4
	[0, 100)	277	277	0	277	0.2	277	14.4	277	0.1	277	0.2
	[100, 1800)	272	272	0.4	272	0.5	272	371.5	272	0.3	272	0.4
	[1800, ∞)	1300	617	250.2	838	134.1	76	3474.1	1087	53.1	1279	35.5
MUCP	All	268	172	162.7	216	70.8	240	131.4	–	–	266	39.0
	[0, 10)	26	26	3.1	26	3.0	26	5.0	–	–	26	3.6
	[10, 300)	101	101	19.2	101	14.1	101	31.3	–	–	101	15.8
	[300, ∞)	34	1	3353.4	7	2789.0	18	1972.0	–	–	32	751.3

geometric mean, with a shift of 1 s. For instances that are not solved to optimality within the time limit, the solving time is set to 3600 s.

In Table 1, we see that over all MKP instances, the activation handler method solves more instances to optimality within the time limit, compared to the other models. We can see that symmetry handling is highly relevant for the MKP problem. SCIP's state-of-the-art symmetry-handling methods reduce the running time by roughly 33%. Our activation handler approach reduces the running time of this already very competitive setting by further 63%. Comparing the different sub-symmetry-handling methods, the additional overhead necessary for the inequalities method is too large for handling this type of sub-symmetries. From the small and medium classes, we see there is a substantial difference between solving time for the inequalities method. There is also a slight improvement in running time for the global orbitope and activation handler methods, compared to default SCIP and the model where no symmetry handling is performed. For the large instances we see that the activation handler method solves more instances to optimality with a clear improvement in solving time.

For the MUCP we see similar results. Overall, SCIP's symmetry-handling methods improve the running time by roughly 55%, whereas the activation handler reduces it even further by 45%. We omit the Orbitope model in the results, as SCIP automatically finds these orbitopes in the Default model. The inequalities method, in contrast with the results for the MKP, shows improvement on the default models for the large instances, confirming the results of [2]. This difference compared to the results for the MKP is likely caused by the auxiliary z-variables, that can here be expressed linearly with no additional constraints. The activation handler outperforms all other models; it solves considerably more large instances to optimality.

6 Conclusion

Based on our numerical experiments, we can answer our research question to the affirmative, i.e., activation handlers substantially improve on the SHI-based approach. We thus believe that this approach can also be successful in further applications. It is up to future research to investigate this in more detail. To this end, we aim to include our framework in the SCIP solver. This way, also other researchers can easily benefit from our flexible framework and extend to a broader class of problems. Our new callback also enables the exploration of different symmetry handling methods that, e.g., do not depend on a fixed variable order. To the best of our knowledge, it is currently not possible to implement such methods directly in commercial solvers.

Acknowledgements We thank the authors of [2] for providing us with the MUCP instances originally used in their experiments.

References

1. Babai, L., Luks, E.M.: Canonical labeling of graphs. In: Proceedings of the Fifteenth Annual ACM Symposium on Theory of Computing—STOC'83. ACM Press (1983)
2. Bendotti, P., Fouilhoux, P., Rottner, C.: Symmetry-breaking inequalities for ILP with structured sub-symmetry. Math. Program. **183**(1), 61–103 (2020)
3. Bendotti, P., Fouilhoux, P., Rottner, C.: Orbitopal fixing for the full (sub-)orbitope and application to the unit commitment problem. Math. Program. **186**(1), 337–372 (2021)
4. van Doornmalen, J., Hojny, C.: Efficient propagation techniques for handling cyclic symmetries in binary programs (2022). https://optimization-online.org/2022/03/8812/
5. Fischetti, M., Liberti, L.: Orbital shrinking. In: Mahjoub, A.R., Markakis, V., Milis, I., Paschos, V.T. (eds.) Combinatorial Optimization, LNCS, vol. 7422, pp. 48–58. Springer, Berlin (2012)
6. Friedman, E.J.: Fundamental domains for integer programs with symmetries. In: Dress, A., Xu, Y., Zhu, B. (eds.) Combinatorial Optimization and Applications. Lecture Notes in Computer Science, vol. 4616, pp. 146–153. Springer, Berlin, Heidelberg (2007)
7. Fukunaga, A.S.: A branch-and-bound algorithm for hard multiple knapsack problems. Ann. Oper. Res. **184**(1), 97–119 (2011)
8. Garey, M.R., Johnson, D.S.: Computers and Intractability: A Guide to the Theory of NP-Completeness. W. H. Freeman & Co (1979)
9. Hojny, C.: Packing, partitioning, and covering symresearch. Disc. Appl. Math. **283**, 689–717 (2020)
10. Hojny, C., Pfetsch, M.E.: Polytopes associated with symmetry handling. Math. Program. **175**(1), 197–240 (2019)
11. Kaibel, V., Peinhardt, M., Pfetsch, M.E.: Orbitopal fixing. Discr. Optim. **8**(4), 595–610 (2011)
12. Kaibel, V., Pfetsch, M.E.: Packing and partitioning orbitopes. Math. Program. **114**(1), 1–36 (2008)
13. Liberti, L.: Automatic generation of symmetry-breaking constraints. In: Combinatorial Optimization and Applications. LNCS, vol. 5165, pp. 328–338. Springer, Berlin (2008)
14. Liberti, L.: Reformulations in mathematical programming: automatic symmetry detection and exploitation. Math. Program. **131**(1–2), 273–304 (2012)
15. Liberti, L.: Symmetry in mathematical programming. In: Lee, J., Leyffer, S. (eds.) Mixed Integer Nonlinear Programming, IMA Series, vol. 154, pp. 236–286. Springer, New York (2012)

16. Liberti, L., Ostrowski, J.: Stabilizer-based symmetry breaking constraints for mathematical programs. J. Glob. Optim. **60**, 183–194 (2014)
17. Margot, F.: Pruning by isomorphism in branch-and-cut. Math. Program. **94**(1), 71–90 (2002)
18. Margot, F.: Exploiting orbits in symmetric ILP. Math. Program. **98**(1), 3–21 (2003)
19. Margot, F.: Symmetry in Integer Linear Programming, Chap. 17, pp. 647–686. Springer (2010)
20. Martello, S., Toth, P.: Algorithms for knapsack problems. In: Martello, S., Laporte, G., Minoux, M., Ribeiro, C. (eds.) Surveys in Combinatorial Optimization. North-Holland Mathematics Studies, vol. 132, pp. 213–257. North-Holland (1987)
21. Ostrowski, J.: Symmetry in integer programming. Ph.D. dissertation, Lehigh University (2009)
22. Ostrowski, J., Linderoth, J., Rossi, F., Smriglio, S.: Orbital branching. Math. Program. **126**(1), 147–178 (2011)
23. Pfetsch, M.E., Rehn, T.: A computational comparison of symmetry handling methods for mixed integer programs. Math. Program. Comput. **11**(1), 37–93 (2019)
24. Pisinger, D.: An exact algorithm for large multiple knapsack problems. European J. Oper. Res. **114**(3), 528–541 (1999)
25. Salvagnin, D.: A dominance procedure for integer programming. Master's thesis, University of Padova, Padova, Italy (2005)
26. Salvagnin, D.: Symmetry breaking inequalities from the Schreier-Sims table. In: van Hoeve, W.J. (ed.) CPAIOR, pp. 521–529. Springer (2018)
27. SCIP: SCIP optimization suite. https://www.scipopt.org. Accessed 27 Jan. 2023
28. Wessel, S.: Activation handler (2022). https://github.com/stenwessel/activation-handler
29. Wessel, S.: Handling sub-symmetries in integer linear programming using activation handlers. Master's thesis, Eindhoven University of Technology, Eindhoven (2022)

A Multivariate Complexity Analysis of the Generalized Noah's Ark Problem

Christian Komusiewicz and Jannik T. Schestag

Abstract In the GENERALIZED NOAH'S ARK PROBLEM, one is given a phylogenetic tree on a set of species X and a set of projects for each species. Each project comes with a cost and raises the survival probability of the corresponding species. The aim is to select for each species a conservation project such that the total cost of the selected projects does not exceed some given threshold and that the expected phylogenetic diversity is as large as possible. We study GENERALIZED NOAH'S ARK PROBLEM and some of its special cases with respect to several parameters related to the input structure such as the number of different costs, the number of different survival probabilities, or the number of species, $|X|$.

1 Introduction

The preservation of biological diversity is one of humanity's most critical challenges. To help addressing this challenge in a systematic way, it is useful to quantify or predict the effect of interventions. Here, two questions arise: how to measure biological diversity of ecosystems and how to model the effect of certain actions on the biological diversity of an ecosystem under consideration.

A popular measure to measure the biological diversity of an ecosystem, introduced by Faith [4], is to consider the *phylogenetic diversity* of the species present in that system. Here, the phylogenetic diversity is the sum of evolutionary distances between the species, when their evolution is modeled by an evolutionary (phylogenetic) tree. The tree then not only gives the phylogenetic diversity of the whole species set but also allows to infer the phylogenetic diversity of any subset of these species that would remain after some currently present species are extinct. Now to model the effect of certain actions, a first simple model is that one can afford to protect k species

C. Komusiewicz · J. T. Schestag (✉)
Friedrich-Schiller-Universität Jena, Ernst-Abbe-Platz 2, 07743 Jena, Germany
e-mail: j.t.schestag@uni-jena.de

C. Komusiewicz
e-mail: c.komusiewicz@uni-jena.de

A. Brieden et al. (eds.), *Graphs and Combinatorial Optimization: from Theory to Applications*, AIRO Springer Series 13,
https://doi.org/10.1007/978-3-031-46826-1_9

and that all other species go extinct. Under this model phylogenetic diversity can be efficiently maximized with a simple greedy algorithm [8, 9]. Later, more realistic models where introduced. One step was to model that protecting some species may be more costly than protecting others [10]. Subsequent approaches also allowed to model uncertainty: performing an action does not guarantee the survival of a species but only raises its survival probability [6]. In this model, one now aims to maximize the *expected* phylogenetic diversity. Finally, one may also consider the even more realistic case when for each species, one can choose from a set of actions, each coming with a cost and a resulting survival probability. This model was proposed by Billionnet [12, 13] as GENERALIZED NOAH'S ARK PROBLEM (GNAP).

Introducing cost differences for species protection makes the problem of maximizing phylogenetic diversity NP-hard [10] and thus all of the even richer models are NP-hard as well. Apart from several pseudopolynomial-time algorithms, there is no work that systematically studies which structural properties of the input make the problems tractable. The aim of this work is to fill this gap. More precisely, we study how different parameters related to the input structure influence the algorithmic complexity of GNAP and some of its special cases. In a nutshell, we show the following. First, GNAP can be solved more efficiently when the number of different project costs and survival probabilities is small. Second, while a constant number $|X|$ of species X leads to polynomial-time solvability, algorithms with running time $f(|X| |)| \cdot |\mathcal{I}|^{\mathcal{O}(1)}$ are unlikely to exist. Finally, restricted cases where for example the input tree has height 1 or there is exactly one action for each species which guarantees its survival are much easier than the general problem. We also observe close relations to the MULTIPLE- CHOICE KNAPSACK problem and to a further natural number problem which we call PENALTY- SUM.

Due to lack of space, most of our proofs are deferred to the long version [1].

2 Preliminaries

For values i and j, the *Kronecker-Delta* $\delta_{i=j}$ is 1 if $i = j$, and 0 otherwise. For an integer n, we define $[n] := \{1, \ldots, n\}$ and $[n]_0 := \{0, 1, \ldots, n\}$. A *partition of* N is a family of pairwise disjoint sets $\{N_1, \ldots, N_m\}$ such that $\bigcup_{i=1}^{m} N_i = N$.

A *directed graph* G is a tuple (V, E), where V is called the *set of vertices* of G and E the *set of edges* of G, respectively. An edge $e = (v, u)$ is *incident* with u and v. The *degree of a vertex* v is the number of edges that are incident with v. A *tree* T *with root* r is a directed, cycle-free graph with $r \in V(T)$ where each vertex of T can be reached from r via exactly one path. A vertex v of a tree T is a *leaf* when the degree of v is one. In a rooted tree, the *height of a vertex* v is the distance from the root r to v for each vertex v. The *height* height_T *of a rooted tree* T is the maximal height of one of the vertices of T. For an edge (u, v), we call u *the parent of* v and v *a child of* u. For a vertex v with parent u, the *subtree* T_v *rooted at* v is the connected component containing v in $T - (u, v)$. In the special case that v is the root of T, we define $T_v := T$. For a vertex v with children $w_1, \ldots, w_t$ and

$i \in [t]$, the *i-partial subtree* $T_{v,i}$ *rooted at* v is the connected component containing v in $T_v - (v, w_{i+1}) - \cdots - (v, w_t)$. For a vertex v in a tree T, the *offspring of* v, denoted $\text{off}(v)$, is the set of leaves in T_v. A *star* is a graph $G = (V, E)$ with a vertex $c \in V$ and $E = \{(c, v) \mid v \in V \setminus \{c\}\}$.

A phylogenetic X*-tree* $\mathcal{T} = (V, E, \lambda)$ (in short, X-tree) is a tree T with root r, where $\lambda : E \to \mathbb{N}$ is an edge-weight function and X is the set of leaves of T. In biological applications, the internal vertices of T correspond to hypothetical ancestors of the leaves and $\lambda(e)$ describes the evolutionary distance between the endpoints of e. An X-tree $\mathcal{T}$ is *ultrametric* if there is an integer p such that the sum of the weights of the edges from the root r to x_i equals p for every leaf x_i.

A *project* $p_{i,j}$ is a tuple $(c_{i,j}, w_{i,j}) \in \mathbb{N}_0 \times \mathbb{Q} \cap [0, 1]$, where $c_{i,j}$ is called the *cost* and $w_{i,j}$ the *survival probability* of $p_{i,j}$. For a given X-tree $\mathcal{T}$ and a taxon $x_i \in X$, a *project list* P_i is a tuple $(p_{i,1}, \dots, p_{i,\ell_i})$. As a project with higher cost will only be considered when the survival probability is higher, we assume the costs and survival probabilities to be ordered. That is, $c_{i,j} < c_{i,j+1}$ and $w_{i,j} < w_{i,j+1}$ for every project list P_i and every $j < \ell_i$. An m*-collection of projects* $\mathcal{P}$ is a set of m project lists $\{P_1, \dots, P_m\}$. For a project set S, the *total cost* $\text{Cost}(S)$ *of* S is $\sum_{p_{i,j} \in S} c_{i,j}$.

For a given X-tree $\mathcal{T}$, the *phylogenetic diversity* $PD_{\mathcal{T}}(S)$ *of a set of projects* $S = \{p_{1,j_1}, \dots, p_{|X|,j_{|X|}}\}$ is given by

$$PD_{\mathcal{T}}(S) := \sum_{(u,v) \in E} \lambda((u, v)) \cdot \left(1 - \prod_{x_i \in \text{off}(v)} (1 - w_{i,j_i})\right).$$

The term $\left(1 - \prod_{x_i \in \text{off}(v)} (1 - w_{i,j_i})\right)$ is the likelihood that some offspring of v survives. Thus, $PD_{\mathcal{T}}(S)$ is is the expected total edge-weight of $\mathcal{T}$ when applying S.

To assess the influence of structural properties of the input on the problem complexity, we study the problems in the framework of parameterized complexity. For a detailed introduction to parameterized complexity refer to the standard monographs [15, 16]. We only recall the most important definitions: A parameterized problem with parameter k is *fixed-parameter tractable* (FPT) if every instance $(\mathcal{I}, k)$ can be solved in $f(k) \cdot |\mathcal{I}|^{\mathcal{O}(1)}$ time. A parameterized problem is *slicewice-polynomial* (XP) if every instance can be solved in $|\mathcal{I}| \cdot g(k)$ time. It is widely assumed that problems that are hard for W[1] have no FPT-algorithm.

2.1 *Problem Definitions, Parameters, and Results Overview*

We now state our main problem and the case where each species has two projects.

GENERALIZED NOAH'S ARK PROBLEM (GNAP)
Input: An X-tree $\mathcal{T} = (V, E, \lambda)$, a $|X|$-collection of projects $\mathcal{P}$, and numbers $B \in \mathbb{N}_0$, $D \in \mathbb{Q}_{\geq 0}$.

Question: Is there a set of projects $S = \{p_{1,j_1}, \dots, p_{|X|,j_{|X|}}\}$, one from each project list of $\mathcal{P}$, such that $PD_\mathcal{T}(S) \geq D$ and $\text{Cost}(S) \leq B$?

A project set S is called *a solution for the instance* $\mathcal{I} = (\mathcal{T}, \mathcal{P}, B, D)$.

$a_i \xrightarrow{c_i} b_i$ [2]-NOAH'S ARK PROBLEM ($a_i \xrightarrow{c_i} b_i$ [2] -NAP)

Input: An X-tree $\mathcal{T} = (V, E, \lambda)$, a $|X|$-collection of projects $\mathcal{P}$ in which the project list P_i contains exactly two projects $(0, a_i)$ and (c_i, b_i) for all $i \in [|X|]$, and integers $B \in \mathbb{N}_0$, $D \in \mathbb{Q}_{\geq 0}$.
Question: Is there a set of projects $S = \{p_{1,j_1}, \dots, p_{|X|,j_{|X|}}\}$, one from each project list of $\mathcal{P}$, such that $PD_\mathcal{T}(S) \geq D$, and $\text{Cost}(S) \leq B$?

In other words, in an instance $a_i \xrightarrow{c_i} b_i$ [2]-NAP we can decide for each taxon x_i whether we want to spend c_i to increase the survival probability of x_i from a_i to b_i.

We study GNAP with respect to several parameters which we describe in the following; for an overview of the results see Table 1. If not stated differently, we assume in the following that $i \in [|X|]$ and $j \in [|P_i|]$. The input of GNAP directly gives the natural parameters *number of taxa* $|X|$, *budget* B, and required *diversity* D. Closely related to B is the *maximum cost per project* $C = \max_{i,j} c_{i,j}$. We may assume that no projects have a cost that exceeds the budget, as we can delete them from the input and so $C \leq B$. We may further assume that $B \leq C \cdot |X|$, as otherwise we can compute in polynomial time whether the diversity of the most valuable projects of the taxa exceeds D and return yes, if it does and no, otherwise.

Further, we consider the *maximum number of projects per taxon* $L := \max_i |P_i|$. By definition, $L = 2$ in $a_i \xrightarrow{c_i} b_i$ [2]-NAP and in GNAP we have $L \leq C + 1$. We denote the *number of projects* by $\|\mathcal{P}\| = \sum_i |P_i|$. Clearly, $|X| \leq \|\mathcal{P}\|$, $L \leq \|\mathcal{P}\|$, and $\|\mathcal{P}\| \leq |X| \cdot L$. By var_c, we denote the *number of different costs*, that is, $\text{var}_c := |\{c_{i,j} : (c_{i,j}, w_{i,j}) \in P_i, P_i \in \mathcal{P}\}|$. We define the *number of different survival probabilities* var_w analogously. The consideration of this type of parameterization, called *number of numbers* parameterization was initiated by Fellows et al. [11]; it is motivated by the idea that often the number of numbers may be small. Also, we consider the *maximum encoding length for survival probabilities* w-code $= \max_{i,j}($ binary length of $w_{i,j})$ and the *maximum edge weight* $\text{val}_\lambda = \max_{e \in E} \lambda(e)$. Observe that because the maximal survival probability of a taxon could be smaller than 1, one can not assume that $\text{val}_\lambda \leq D$.

2.2 Observations for GNAP

We first present some basic observations that provide some first complexity classifications. In the problem with exactly two projects per taxa, $a_i \xrightarrow{c_i} b_i$ [2]-NAP, one can iterate over all subsets X' of taxa and check if it is a possible solution pay c_i to increase the survival probability for each $x_i \in X'$. To this end, we check if $\sum_{x_i \in X'} c_i \leq B$ and compute if the phylogenetic diversity is at least D, when the survival probability

Table 1 Complexity results for GENERALIZED NOAH'S ARK PROBLEM. Here, $0 \xrightarrow{c_i} 1$ [2]-NAP is the special case where the survival probabilities are only 0 or 1, and $0 \xrightarrow{1} b_i$ [2]-NAP is the special case where each project has unit costs. Entries with the sign "—" mark parameters that are (partially) constant in the specific problem definition and thus are not interesting

Parameter	GNAP	GNAP with $\text{height}_{\mathcal{T}} = 1$
$\lvert X \rvert$	W[1]-hard (Theorem 4.4), XP (Proposition 4.1)	W[1]-hard (Theorem 4.4), XP (Proposition 4.1)
B	XP (Observation 2.2)	PFPT $\mathcal{O}(B \cdot \lVert \mathcal{P} \rVert)$ (Proposition 4.7)
C	Open	PFPT $\mathcal{O}(C \cdot \lVert \mathcal{P} \rVert \cdot \lvert X \rvert)$ (Proposition 4.7)
D	Para-NP-h (Observation 2.4)	Para-NP-h (Observation 2.4)
val_λ	Para-NP-h (Theorem 4.4)	Para-NP-h (Theorem 4.4)
var_c	Open	XP $\mathcal{O}(\lvert X \rvert^{\text{var}_c - 1} \cdot \lVert \mathcal{P} \rVert)$ (Proposition 4.7)
var_w	Para-NP-h (Observation 2.3)	Para-NP-h (Observation 2.3)
D + w-code	Open	FPT $\mathcal{O}(D \cdot 2^{\text{w-code}} \cdot \lVert \mathcal{P} \rVert)$ (Proposition 4.7)
$B + \text{var}_w$	XP $\mathcal{O}(B \cdot \lvert X \rvert^{2 \cdot \text{var}_w + 1})$ (Theorem 4.3)	PFPT (Proposition 4.7)
$D + \text{var}_w$	Para-NP-h (Observation 2.4)	Para-NP-h (Observation 2.4)
$\text{var}_c + \text{var}_w$	XP $\mathcal{O}(\lvert X \rvert^{2 \cdot (\text{var}_c + \text{var}_w) + 1})$ (Theorem 4.2)	FPT (Theorem 4.8)
Parameter	$0 \xrightarrow{c_i} 1$ [2]-NAP	$0 \xrightarrow{1} b_i$ [2]-NAP
$\lvert X \rvert$	FPT (Observation 2.1)	FPT (Observation 2.1)
B	PFPT $\mathcal{O}(B^2 \cdot n)$ [10]	XP (Observation 2.2)
C	PFPT $\mathcal{O}(C^2 \cdot n^3)$ (Corollary 5.1)	—
D	PFPT $\mathcal{O}(D^2 \cdot n)$ (Proposition 5.2)	Open
val_λ	PFPT $\mathcal{O}((\text{val}_\lambda)^2 \cdot n^3)$ [1]	Open
var_c	XP [1]	—
var_w	—	XP [1]
D + w-code	—	Open
$B + \text{var}_w$	—	XP [1]
$D + \text{var}_w$	—	Open

of every $x_i \in X'$ is b_i and a_i otherwise. Thus, $a_i \xrightarrow{c_i} b_i$ [2]-NAP is fixed-parameter tractable with respect to the number of taxa.

Observation 2.1 *$a_i \xrightarrow{c_i} b_i$ [2]-NAP can be solved in $2^{|X|} \cdot |\mathcal{I}|^{\mathcal{O}(1)}$ time.*

A GNAP solution contains at most B projects with positive costs. Hence, a solution can be found by iterating over all B-sized subsets X' of taxa and checking every combination of projects for X'. Like before, we have to check that the budget is not exceeded and the phylogenetic diversity of the selected projects is at least D. This brute-force algorithm shows that GNAP is XP with respect to the budget.

Observation 2.2 *GNAP can be solved in* $(|X| \cdot L)^B \cdot |\mathcal{I}|^{\mathcal{O}(1)}$ *time.*

In the NP-hard KNAPSACK problem, one is given a set of items N, a cost-function $c : N \to \mathbb{N}$, a value-function $d : N \to \mathbb{N}$, and two integers B and D and asks whether there is a an item set N' such that $c(N') \leq B$ and $d(N') \geq D$. We describe briefly a known reduction from KNAPSACK to $0 \overset{c_i}{\to} 1$ [2]-NAP [10]. Let $\mathcal{I} = (N, c, d, B, D)$ be an instance of KNAPSACK. Define $\mathcal{T} := (V, E, \lambda)$ to be an N-tree with $V := \{w\} \cup N$ and $E := \{(w, x_i) \mid x_i \in N\}$ and $\lambda((w, x_i)) := d(x_i)$. For each leaf x_i we define a project list P_i that contains two projects $(0, 0)$ and $(c(x_i), 1)$. Then, $\mathcal{I}' := (\mathcal{T}, \mathcal{P}, B' := B, D' := D)$ is a yes-instance of $0 \overset{c_i}{\to} 1$ [2]-NAP if and only if $\mathcal{I}$ is a yes-instance of KNAPSACK.

Observation 2.3 ([10]) $0 \overset{c_i}{\to} 1$ [2]-NAP *is* NP-*hard, even if the tree* $\mathcal{T}$ *has height 1.*

Because $0 \overset{c_i}{\to} 1$ [2]-NAP is a special case of GNAP in which $L = 2$, w-code $= 1$, and $\text{var}_w = 2$, we conclude that GNAP is NP-hard, even if $\text{height}_\mathcal{T} = \text{w-code} = 1$ and $L = \text{var}_w = 2$. In this reduction, one could also set $D' := 1$ and set the survival probability of every project with positive cost to $b := 1/D$.

Observation 2.4 $0 \overset{c_i}{\to} b$ [2]-NAP *is* NP-*hard, even if* $D = 1$, $b \in (0, 1]$ *is a constant, and the given* X*-tree* $\mathcal{T}$ *has height 1.*

Thus, GNAP is NP-hard even if $D = 1$ and $\text{var}_w = 2$. This however is, because the size of the binary encoding of a survival probability became very large. Thus, one can wonder if GNAP admits an FPT algorithm for the parameter $D +$ w-code. Proposition 4.7 shows such an algorithm for the case when $\mathcal{T}$ has $\text{height}_\mathcal{T} = 1$.

3 Multiple-Choice Knapsack

In this section, we consider a variant of KNAPSACK, in which the set of items is divided into classes. From every class, exactly one item can be chosen.

MULTIPLE- CHOICE KNAPSACK PROBLEM (MCKP)
Input: A set of items $N = \{a_1, \ldots, a_n\}$, a partition $\{N_1, \ldots, N_m\}$ of N, two functions $c, d : N \to \mathbb{N}$, and two integers B, D.
Question: Is there a set $S \subseteq N$ such that $|S \cap N_i| = 1$ for each $i \in [m]$, $c_\Sigma(S) \leq B$, and $d_\Sigma(S) \geq D$?

Herein, we write $c_\Sigma(A) := \sum_{a_i \in A} c(a_i)$ and $d_\Sigma(A) := \sum_{a_i \in A} d(a_i)$ for a set $A \subseteq N$. We call $c(a_i)$ the *cost of* a_i and $d(a_i)$ the *value of* a_i. Further, for a set $A \subseteq N$ we define $c(A) := \{c(a) \mid a \in A\}$ and $d(A) := \{d(a) \mid a \in A\}$. A set S that fulfills the presented criteria is called a *solution* for the instance $\mathcal{I}$.

For MCKP, we consider parameters that are closely related to the parameters described for GNAP: The input gives the *number of classes* m, the *budget* B, and the desired *value* D. Closely related to B is the *maximum cost for an item* $C =$

Table 2 Complexity results for MULTIPLE- CHOICE KNAPSACK

Parameter	MCKP
m	W[1]-hard, XP (Theorem 3.6)
B	PFPT $\mathcal{O}(B \cdot \|N\|)$ [5]
C	PFPT $\mathcal{O}(C \cdot \|N\| \cdot m)$ (Observation 3.1)
D	PFPT $\mathcal{O}(D \cdot \|N\|)$ [7]
L	para-NP [17]
var_c	XP $\mathcal{O}(m^{\text{var}_c - 1} \cdot \|N\|)$ (Proposition 3.2)
var_d	XP $\mathcal{O}(m^{\text{var}_d - 1} \cdot \|N\|)$ (Proposition 3.3)
$\text{var}_c + \text{var}_d$	FPT (Theorem 3.4)

$\max_{a_j \in N} c(a_j)$. As for GNAP, we may assume $C \leq B$ and $B \leq C \cdot m$. By var_c, we denote the *number of different costs*, that is, $\text{var}_c := |\{c(a_j) : a_j \in N\}|$. We define the *number of different values* var_d analogously. The size of the biggest class is denoted by L. If one class N_i contains two items a_p and a_q with the same cost and $d(a_p) \leq d(a_p)$, the item a_p can be removed from the instance. Thus, we may assume that no class contains two items with the same cost and so $L \leq \text{var}_c$. Analogously, we may assume that no class contains two same-valued items and so $L \leq \text{var}_w$. Table 2 lists old and new complexity results for MCKP.

First, we provide some algorithms for MCKP. It is known that MCKP can be solved in $\mathcal{O}(B \cdot |N|)$ time [5] and in $\mathcal{O}(D \cdot |N|)$ time [7]. As we may assume that $C \cdot m \geq B$, we may also observe the following.

Observation 3.1 MCKP *can be solved in* $\mathcal{O}(C \cdot |N| \cdot m)$ *time.*

KNAPSACK is FPT with respect to the number of different costs var_c [14], via reduction to ILP- FEASIBILITY with $f(\text{var}_c)$ variables. This approach can not be adopted easily, as it has to be checked whether a solution contains exactly one item per class. In Propositions 3.2 and 3.3 we show that MCKP is XP with respect to the number of different costs and different values, respectively. Then, in Theorem 3.4 we show that MCKP is FPT with respect to the parameter $\text{var}_c + \text{var}_d$.

Proposition 3.2 *MCKP can be solved in* $\mathcal{O}(m^{\text{var}_c - 1} \cdot |N|)$ *time, where* var_c *is the number of different costs.*

Proposition 3.3 *MCKP can be solved in* $\mathcal{O}(m^{\text{var}_d - 1} \cdot |N|)$*, where* var_d *is the number of different values.*

By Propositions 3.2 and 3.3, MCKP is XP with respect to var_c and var_d, respectively. In the following, we reduce an instance of MCKP to an instance of ILP- FEASIBILITY, in which the number of variables is in $2^{\text{var}_c + \text{var}_d} \cdot \text{var}_c$. Since ILP- FEASIBILITY with n variables and input length s can be solved using $s \cdot n^{2.5n + o(n)}$ arithmetic operations [2, 3], this reduction gives the following.

Theorem 3.4 *For an instance of MCKP an equivalent instance of* ILP- FEASIBILITY *with* $\mathcal{O}(2^{\text{var}_c + \text{var}_d} \cdot \text{var}_c)$ *variables can be defined. Thus, MCKP is* FPT *with respect to* $\text{var}_c + \text{var}_d$.

We contrast these algorithms by the following hardness results. There is a reduction from KNAPSACK to MCKP in which each item in the instance of KNAPSACK is added to a unique class with a further, new item that has no costs and no value [17].

Observation 3.5 *([17])* MCKP *is* NP-*hard even if every class contains two items.*

In the following, we prove that MCKP is W[1]-hard with respect to the number of classes m, even if $B = D$ and $c(a) = d(a)$ for each $a \in N$. This special case of MCKP is called MULTIPLE- CHOICE SUBSET SUM [17] chosen.

Theorem 3.6 MCKP *is* XP *and* W[1]-*hard with respect to the number of classes* m.

4 The Generalized Noah's Ark Problem

We now consider the Generalized Noah's Ark Problem (GNAP). First, we observe that for a constant number of taxa, we can solve the problem in polynomial time by branching into the possible project choices for each taxon.

Proposition 4.1 GNAP *is* XP *with respect to* $|X|$.

In Theorem 4.2, we now show that GNAP can be solved in polynomial time when the number of different project costs and the number of different survival probabilities is constant.

In the following, let $\mathcal{I} = (\mathcal{T}, \lambda, \mathcal{P}, B, D)$ be an instance of GNAP, and let $\mathcal{C} := \{c_1, \ldots, c_{\text{var}_c}\}$ and $\mathcal{W} := \{w_1, \ldots, w_{\text{var}_w}\}$ denote the sets of different costs and different survival probabilities in $\mathcal{I}$, respectively. Without loss of generality, assume $c_i < c_{i+1}$ for each $i \in [\text{var}_c - 1]$ and likewise assume $w_j < w_{j+1}$ for each $j \in [\text{var}_w - 1]$. In other words, c_i is the ith cheapest cost in $\mathcal{C}$ and w_j is the jth smallest survival probability in $\mathcal{D}$. Recall, that we assume that there is at most one item with cost c_p and at most one item with survival probability w_q in every project list P_i, for each $p \in [\text{var}_c]$ and $q \in [\text{var}_w]$. For the rest of the section, by $\mathbf{a}$ and $\mathbf{b}$ we denote $(a_1, \ldots, a_{\text{var}_c - 1})$ and $(b_1, \ldots, b_{\text{var}_w - 1})$, respectively. Further, we let $\mathbf{p}_{(j)+z}$ denote the vector $\mathbf{p}$ in which at position i, the value z is added and we let $\mathbf{0}$ denote the $(\text{var}_c - 1)$-dimensional zero.

Theorem 4.2 GNAP *can be solved in* $\mathcal{O}\left(|X|^{2(\text{var}_c + \text{var}_w - 1)} \cdot (\text{var}_c + \text{var}_w)\right)$ *time.*

Proof *Algorithm* We describe a dynamic programming algorithm with two tables F and G that have a dimension for all the var_c different costs, except for c_{var_c} and all the var_w different survival probabilities, except for $\text{var}_w - 1$. Recall that for a vertex v with t children T_v is the subtree rooted at v and the offspring $\text{off}(v)$ of v are the leaves in T_v. We define the *i-partial subtree* $T_{v,i}$ rooted at v as the subtree of T_v containing only the first i children of v for $i \in [t]$. For a vertex $v \in V$ and given vectors $\mathbf{a}$ and $\mathbf{b}$, we define $\mathcal{S}_{\mathbf{a},\mathbf{b}}^{(v)}$ to be the family of sets of projects S such that

1. S contains exactly one project of P_i for each $x_i \in \text{off}(v)$, and
2. S contains exactly a_k projects with cost c_k for each $k \in [\text{var}_c - 1]$, and
3. S contains exactly b_ℓ projects with survival probability w_ℓ for each $\ell \in [\text{var}_w - 1]$.

For a vector $v \in V$ with children $u_1, \ldots, u_t$, given vectors $\mathbf{a}$ and $\mathbf{b}$ and a given integer $i \in [t]$ we define $\mathcal{S}_{\mathbf{a},\mathbf{b}}^{(v,i)}$ analogously, just that exactly one projects of P_i is chosen for each $x_i \in \text{off}(u_1) \cup \cdots \cup \text{off}(u_i)$.

It follows that we can compute how many projects with cost c_{var_c} and survival probability w_{var_w} a set $S \in \mathcal{S}_{\mathbf{a},\mathbf{b}}^{(v)}$ contains. That are $a_{\text{var}_c}^{(v)} := |\text{off}(v)| - \sum_{j=1}^{\text{var}_c - 1} a_j$ projects with cost c_{var_c} and $b_{\text{var}_w}^{(v)} := |\text{off}(v)| - \sum_{j=1}^{\text{var}_w - 1} b_j$ projects with survival probability w_{var_w}. The entries $F[v, \mathbf{a}, \mathbf{b}]$ and $G[v, i, \mathbf{a}, \mathbf{b}]$ store the maximum expected phylogenetic diversity of the tree T_v for $S \in \mathcal{S}_{\mathbf{a},\mathbf{b}}^{(v)}$ and $T_{v,i}$ for $S \in \mathcal{S}_{\mathbf{a},\mathbf{b}}^{(v,i)}$, respectively. We further define the total survival probability to be $w(b_{\text{var}_w}, \mathbf{b}) := 1 - (1 - w_{\text{var}_w})^{b_{\text{var}_w}} \cdot \prod_{i=1}^{\text{var}_w - 1} (1 - w_i)^{b_i}$, when b_{var_w} and $\mathbf{b}$ describe the number of chosen single survival probabilities.

Fix a taxon x_i with project list P_i. As we want to select exactly one project of P_i, the project is clearly defined by $\mathbf{a}$ and $\mathbf{b}$. So, we store $F[x_i, \mathbf{a}, \mathbf{b}] = 0$, if P_i contains a project $p = (c_k, w_\ell)$ such that

1. ($k < \text{var}_c$ and $\mathbf{a} = \mathbf{0}_{(k)+1}$ or $k = \text{var}_c$ and $\mathbf{a} = \mathbf{0}$), and
2. ($\ell < \text{var}_w$ and $\mathbf{b} = \mathbf{0}_{(\ell)+1}$ or $\ell = \text{var}_w$ and $\mathbf{b} = \mathbf{0}$).

Otherwise, store $F[x_i, \mathbf{a}, \mathbf{b}] = -\infty$.

Let v be an internal vertex with children $u_1, \ldots, u_t$, we define

$$G[v, 1, \mathbf{a}, \mathbf{b}] = F[u_1, \mathbf{a}, \mathbf{b}] + \lambda((v, u_1)) \cdot w\left(b_{\text{var}_w}^{(u_1)}, \mathbf{b}\right) \tag{1}$$

and to compute further values of G, we can use the recurrence

$$G[v, i+1, \mathbf{a}, \mathbf{b}] = \max_{\substack{\mathbf{0} \le \mathbf{a}' \le \mathbf{a} \\ \mathbf{0} \le \mathbf{b}' \le \mathbf{b}}} \begin{cases} G[v, i, \mathbf{a} - \mathbf{a}', \mathbf{b} - \mathbf{b}'] + F[u_{i+1}, \mathbf{a}', \mathbf{b}'] \\ +\lambda((v, u_{i+1})) \cdot w\left(b_{\text{var}_w}^{(u_{i+1})}, \mathbf{b}'\right) \end{cases}. \tag{2}$$

Herein, we write $\mathbf{p} \le \mathbf{q}$ if $\mathbf{p}$ and $\mathbf{q}$ have the same dimension d and $p_i \le q_i$ for every $i \in [d]$. And finally, we define $F[v, \mathbf{a}, \mathbf{b}] = G[v, t, \mathbf{a}, \mathbf{b}]$.

Return yes if there are $\mathbf{a}$ and $\mathbf{b}$ such that $\sum_{i=1}^{\text{var}_c - 1} a_i \le |X|$, and $\sum_{i=1}^{\text{var}_w - 1} b_i \le |X|$, and $a_{\text{var}_c}^{(r)} \cdot c_{\text{var}_c} + \sum_{i=1}^{\text{var}_c - 1} a_i \cdot c_i \le B$, and $F[r, \mathbf{a}, \mathbf{b}] \ge D$ where r is the root of $\mathcal{T}$. Otherwise, return no.

The correctness and running time proofs are deferred to the long version [1]. □

As each project with a cost higher than B can be deleted, we may assume that there are no such projects which implies that $\text{var}_c \le C + 1 \le B + 1$. Thus, Theorem 4.2 also implies that GNAP is XP with respect to $C + \text{var}_w$ and $B + \text{var}_w$ with an astronomical running time of $\mathcal{O}\left(|X|^{2(C+\text{var}_w - 1)} \cdot (C + \text{var}_w)\right)$ and $\mathcal{O}\left(|X|^{2(B+\text{var}_w - 1)} \cdot (B + \text{var}_w)\right)$, respectively. However, however we can adjust algorithm so that B is not in the exponent of the running time. Instead of declaring how many projects of cost c_i for $i \in [\mathrm{var}_c]$ are selected, we declare the budget that can be spent.

Theorem 4.3 GNAP *can be solved in* $\mathcal{O}\left(B^2 \cdot |X|^{2(\mathrm{var}_w - 1))} \cdot \mathrm{var}_w\right)$ *time.*

We now consider the special case of GNAP where the X-tree $\mathcal{T}$ has height 1. We first show that this special case—and therefore GNAP—is W[1]-hard with respect to the number $|X|$ of taxa. This implies that Proposition 4.1 cannot be improved to an FPT algorithm. Afterward, we prove that most of the FPT and XP algorithms we presented for MCKP can also be adopted for this special case of GNAP yielding algorithms that have a faster running time than for GNAP. Recall that Δ is the highest degree of a vertex in the tree.

Theorem 4.4 GNAP *is* W[1]*-hard with respect to* $|X| + \Delta$*, even if the given* X*-tree* $\mathcal{T}$ *is ultrametric with* $\mathrm{val}_\lambda = \mathrm{height}_\mathcal{T} = 1$*, and* $D = 1$*.*

Proof *Reduction.* We reduce from MCKP, which by Theorem 3.6 is W[1]-hard with respect to the number of classes m. Let $\mathcal{I} = (N, \{N_1, \ldots, N_m\}, c, d, B, D)$ be an instance of MCKP. We define an instance $\mathcal{I}' = (\mathcal{T}, \mathcal{P}, B' := B, D' := 1)$ in which the X-tree $\mathcal{T} = (V, E, \lambda)$ is a star with center v and the vertex set is $V := \{v\} \cup X$, with $X := \{x_1, \ldots, x_m\}$. Set $\lambda(e) := 1$ for every $e \in E$. For every class $N_i = \{a_{i,1}, \ldots, a_{i,\ell_i}\}$, define a project list P_i with projects $p_{i,j} := (c_{i,j} := c(a_{i,j}), w_{i,j} := d(a_{i,j})/D)$. The $|X|$-collection of projects $\mathcal{P}$ contains all these project lists P_i.

Correctness. Because we may assume that $0 \le d(a) \le D$ for all $a \in N$, the survival probabilities fulfill $w_{i,j} \in [0, 1]$ for all $i \in [m]$ and $j \in [|N_i|]$. The tree has m taxa and a maximum degree of m. The reduction is clearly computable in polynomial time, so it only remains to show the equivalence.

"$\Rightarrow$": Let S be a solution for $\mathcal{I}$ with $S \cap N_i = \{a_{i,j_i}\}$. We show that $S' = \{p_{i,j_i} \mid i \in [m]\}$ is a solution for $\mathcal{I}'$: The cost of the set S' is $\sum_{i=1}^m c_{i,j_i} = \sum_{i=1}^m c(a_{i,j_i}) \le B$ and further $PD_\mathcal{T}(S') = \sum_{(v,x_i)\in E} \lambda((v, x_i)) \cdot w_{i,j_i} = \sum_{(v,x_i)\in E} 1 \cdot d(a_{i,j})/D = \frac{1}{D} \cdot \sum_{i=1}^m d(a_{i,j}) \ge 1 = D'$.

"$\Leftarrow$": Let $S = \{p_{1,i_1}, \ldots, p_{m,j_m}\}$ be a solution for $\mathcal{I}'$. We show that $S' = \{a_{1,i_1}, \ldots, a_{m,j_m}\}$ is a solution for $\mathcal{I}$: Clearly, S' contains exactly one item per class. The cost of the set S' is $c(S') = \sum_{i=1}^m c(a_{i,j_i}) = \sum_{i=1}^m c_{i,j_i} \le B$. The diversity of the set S' is $d(S') = \sum_{i=1}^m d(a_{i,j_i}) = \sum_{i=1}^m w_{i,j_i} \cdot D = PD_\mathcal{T}(S) \cdot D \ge D$. □

By Observation 3.5, MCKP is NP-hard, even if every class contains at most two items (of which one has no cost and no value). Because the above reduction is computed in polynomial time, we conclude the following.

Corollary 4.5 $0 \xrightarrow{c_i} b_i$ [2]-NAP *is* NP*-hard, even if the given* X*-tree* $\mathcal{T}$ *is ultrametric with* $\mathrm{height}_\mathcal{T} = \mathrm{val}_\lambda = 1$*, and* $D = 1$*.*

The X-tree that has been constructed in the reduction in the proof of Theorem 4.4, is a star and therefore has a relatively high degree. In the following, we show that GNAP is also W[1]-hard with respect to $|X|$ when $\Delta = 3$.

Corollary 4.6 GNAP *is* W[1]*-hard with respect to* $|X| + D + \text{height}_{\mathcal{T}}$ *even if* $\Delta = 3$ *and* $\text{val}_\lambda = 1$.

In Sect. 3, we presented algorithms that solve MCKP. Many of these algorithms can be adopted for the special case of GNAP where the input X-tree $\mathcal{T}$ has height 1.

Proposition 4.7 *When the given* X*-tree* $\mathcal{T}$ *has* $\text{height}_{\mathcal{T}} = 1$, GNAP *can be solved*

- *in* $\mathcal{O}(D \cdot 2^{\text{w-code}} \cdot \|\mathcal{P}\| + |\mathcal{I}|)$ *time, or*
- *in* $\mathcal{O}(B \cdot \|\mathcal{P}\| + |\mathcal{I}|)$ *time, or*
- *in* $\mathcal{O}(C \cdot \|\mathcal{P}\| \cdot |X| + |\mathcal{I}|)$ *time, or*
- *in* $\mathcal{O}(|X|^{\text{var}_c - 1} \cdot \|\mathcal{P}\| + |\mathcal{I}|)$ *time,*

where $\|\mathcal{P}\| = \sum_{i=1}^{|X|} |P_i|$ *is the number of projects and* $|\mathcal{I}|$ *is the size of the input.*

By Proposition 4.7 and Theorem 3.4, we can conclude that the restriction of GNAP to instances with height 1 is FPT with respect to $\text{var}_c + \text{var}_w + \text{val}_\lambda$. However, in the following, we present a reduction from an instance of GNAP in which the height of the given tree is 1 to an instance of ILP- Feasibility, in which the number of variables is in $\mathcal{O}(2^{\text{var}_c + \text{var}_d} \cdot \text{var}_c)$.

Theorem 4.8 *For an instance of* GNAP *with an* X*-tree of height 1, an equivalent instance of* ILP- Feasibility *with* $\mathcal{O}(2^{\text{var}_c + \text{var}_d} \cdot \text{var}_c)$ *variables can be defined. Thus,* GNAP *is* FPT *with respect to* $\text{var}_c + \text{var}_d$ *if the height of the given* X*-tree is 1.*

5 Restriction to Two Projects per Taxon

We finally study two special cases of $a_i \xrightarrow{c_i} b_i$ [2]-NAP—the special case of GNAP, in which every project list contains exactly two projects.

First, we consider $0 \xrightarrow{c_i} 1$ [2]-NAP, the special case where each taxon x_i survives definitely if cost c_i is paid and becomes extinct, otherwise. This special case was introduced by Pardi and Goldman [10] who also presented a pseudo-polynomial algorithm which computes a solution to $0 \xrightarrow{c_i} 1$ [2]-NAP in $\mathcal{O}(B^2 \cdot n)$ time. Because we may assume that $B \leq C \cdot |X|$, we may conclude the following.

Corollary 5.1 $0 \xrightarrow{c_i} 1$ [2]-NAP *can be solved in* $\mathcal{O}(C^2 \cdot |X|^3)$ *time.*

We also show that $0 \xrightarrow{c_i} 1$ [2]-NAP is FPT with respect to D, with an adaption of the above-mentioned dynamic programming algorithm of Pardi and Goldman [10] for the parameter B.

Proposition 5.2 $0 \xrightarrow{c_i} 1$ [2]-NAP *can be solved in* $\mathcal{O}(D^2 \cdot n)$ *time.*

Second, we consider $0 \xrightarrow{1} b_i$ [2]-NAP—the special case of GNAP in which every project with a positive survival probability has the same cost. Observe that for every $c \in \mathbb{N}$, an instance $\mathcal{I} = (\mathcal{T}, \mathcal{P}, B, D)$ of $0 \xrightarrow{c} b_i$ [2]-NAP can be reduced to an equivalent instance $\mathcal{I}' = (\mathcal{T}, \mathcal{P}', B', D)$ of $0 \xrightarrow{1} b_i$ [2]-NAP by replacing every project (c, b_i) with $(1, b_i)$, and setting $B' = \lfloor B/c \rfloor$. Thus, $0 \xrightarrow{1} b_i$ [2]-NAP can be considered as the special case of GNAP with unit costs for projects.

Unfortunately, we were not able to resolve whether $0 \xrightarrow{1} b_i$ [2]-NAP is NP-hard or not. However, we may relate its complexity to the following, more basic problem.

PENALTY- SUM
Input: A set of tuples $T = \{t_i = (a_i, b_i) \mid i \in [n], a_i \in \mathbb{Q}_{\geq 0}, b_i \in (0, 1)\}$, two integers k, Q, and a number $D \in \mathbb{Q}_+$.
Question: Is there a subset $S \subseteq T$ of size k such that $\sum_{t_i \in S} a_i - Q \cdot \prod_{t_i \in S} b_i \geq D$?
Despite being quite natural and fundamental, we are not aware of any previous work on the complexity of PENALTY- SUM. We present two karp-reductions, one from PENALTY- SUM to $0 \xrightarrow{1} b_i$ [2]-NAP in which the X-tree has a height of 2 and $\deg(r) = 1$ for the root r and one for the converse direction.

Theorem 5.3 PENALTY- SUM *is NP-hard if and only if* $0 \xrightarrow{1} b_i$ [2]-NAP *restricted to X-trees with height 2 and* $\deg(r) = 1$ *for the root r is NP-hard.*

Recall that in an ultrametric tree, we require the length from the root to a vertex to be the same for all vertices. $0 \xrightarrow{1} b_i$ [2]-NAP can be solved greedily on ultrametric trees that have height at most 2 when always the taxon with the highest diversity is selected. In the following theorem, we show that $0 \xrightarrow{1} b_i$ [2]-NAP is even on ultrametric trees of height 3 NP-hard, if PENALTY- SUM is NP-hard.

Theorem 5.4 $0 \xrightarrow{1} b_i$ [2]-NAP *is NP-hard on ultrametric trees of height 3, if* PENALTY- SUM *is NP-hard.*

6 Discussion

We have provided several tractability and intractability results for GNAP and some of its special cases. Naturally, several open questions remain. For example, it is not known whether GNAP is weakly or strongly NP-hard. Moreover, it remains open whether GNAP is FPT with respect to $\text{var}_c + \text{var}_w$. Finally, as described above, it is open whether $0 \xrightarrow{1} b_i$ [2]-NAP and PENALTY- SUM are NP-hard.

References

1. Komusiewicz, C., Schestag, J.: A multivariate complexity analysis of the generalized Noah's ark problem. arXiv:2307.03518. https://doi.org/10.48550/arXiv.2307.03518
2. Lenstra Jr, H.W.: Integer programming with a fixed number of variables. Math. Oper. Res. **8**(4), 538–548 (1983)
3. Frank, A., Tardos, É.: An application of simultaneous diophantine approximation in combinatorial optimization. Combinatorica **7**(1), 49–65 (1987)
4. Faith, D.P.: Conservation evaluation and phylogenetic diversity. Biol. Cons. **61**(1), 1–10 (1992)
5. Pisinger, D.: A minimal algorithm for the multiple-choice Knapsack Problem. Eur. J. Oper. Res. **83**(2), 394–410 (1995)
6. Weitzman, M.L.: The Noah's ark problem. Econometrica 1279–1298 (1998)
7. Bansal, M., Venkaiah, V.: Improved fully polynomial time approximation scheme for the 0–1 multiple-choice Knapsack problem. In: International Institute of Information Technology Technical Report, pp. 1–9 (2004)
8. Pardi, F., Goldman, N.: Species choice for comparative genomics: being greedy works. PLoS Genet. **1**(6), e71 (2005)
9. Steel, M.: Phylogenetic diversity and the greedy algorithm. Syst. Biol. **54**(4), 527–529 (2005)
10. Pardi, F., Goldman, N.: Resource-aware taxon selection for maximizing phylogenetic diversity. Syst. Biol. **56**(3), 431–444 (2007)
11. Fellows, M.R., Gaspers, S., Rosamond, F.A.: Parameterizing by the number of numbers. Theory Comput. Syst. **50**(4), 675–693 (2012)
12. Billionnet, A.: Solution of the generalized Noah's ark problem. Syst. Biol. **62**(1), 147–156 (2013)
13. Billionnet, A.: How to take into account uncertainty in species extinction probabilities for phylogenetic conservation prioritization. Environ. Model. Assess. **22**(6), 535–548 (2017)
14. Etscheid, M., Kratsch, S., Mnich, M., Röglin, H.: Polynomial kernels for weighted problems. J. Comput. Syst. Sci. **84**, 1–10 (2017)
15. Cygan, M., Fomin, F.V., Kowalik, L., Lokshtanov, D., Marx, D., Pilipczuk, M., Pilipczuk, M., Saurabh, S.: Parameterized Algorithms. Springer, Marcin Pilipczuk (2015)
16. Downey, R.G., Fellows, M.R.: Fundamentals of Parameterized Complexity. In: Texts in Computer Science. Springer, Berlin (2013)
17. Kellerer, H., Pferschy, U., Pisinger, D.: Knapsack Problems. Springer, Berlin (2004)

Comparing Ad-Hoc and MIP-Based Algorithms for the Online Facility Location Problem

Rosario Messana and Alberto Ceselli

Abstract We consider online variants of the uncapacitated facility location problem. Facilities need to be placed, at a cost, and clients need to be assigned to them, yielding revenues. We provide an experimental comparison of two classes of algorithms: ad-hoc ones, which rely on the specific structure of the problem, and generic ones, which rely on the solution of Mixed Integer Programs (MIPs) as sub-problems. Models and algorithms from the literature assume one client to appear at a time. We generalize them, assuming that clients may arrive in batches of fixed (but arbitrary) size. We compare our batch adaptation to the original versions of the algorithms. We design four generators of rewards and costs, two being "adversarial" and two stochastic. We propose a variant of an existent MIP-based algorithm to profitably deal with stochastic settings. Our analysis shows that in each of the four settings, suitable MIP-based algorithms provide better solutions than ad-hoc ones, with a comparable computing effort. Our experiments also show that batching is in fact useful.

1 Introduction

In the classical uncapacitated facility location problem (FLP) a set of candidate location sites and a set of clients are given. The decision maker needs to open facilities in a subset of the candidate sites, at a given cost. Each client needs to be assigned to one of the open facilities, yielding a given revenue. Besides its theoretical interest, the FLP contains in its simplicity the core complexity of location problems. It is in fact NP-hard to find a set of sites where facilities need to be opened, which maximizes the difference between the assignment revenues and the opening costs.

The classical FLP, as an offline optimization problem, has been investigated for decades [5]. The FLP appears also in *online* variants, although the literature on them

R. Messana (✉) · A. Ceselli
Dipartimento di Informatica, Universitá degli Studi di Milano, Milan, Italy
e-mail: rosario.messana@unimi.it

A. Ceselli
e-mail: alberto.ceselli@unimi.it

A. Brieden et al. (eds.), *Graphs and Combinatorial Optimization: from Theory to Applications*, AIRO Springer Series 13,
https://doi.org/10.1007/978-3-031-46826-1_10

is much more recent and scarce [13]. In these online FLPs the decision process is iterative. At each iteration, *after* the facilities have been placed, the location costs and the assignment rewards are revealed. One of the most studied variants assumes that a single client arrives at each round. The decision maker is allowed to open an arbitrary number of facilities, paying their costs; only the revenue of the best assignment is however collected. Recent contributions include [19], in which the authors propose an algorithm named MaxHedge, and carry out a theoretical analysis of its quality guarantees. Being the online counterpart of the FLP, generic online optimization techniques relying on solving a MIP subproblem at each round can also be exploited [22]. Despite the relevance of FLP in applications [5], to the best of our knowledge, algorithm [19] has not been experimented in practice nor compared to algorithms like [22].

A first research question therefore arises, on whether an ad-hoc approach like [19] is actually yielding advantages with respect to a generic MIP-based approach like that of [22]. Concerning modeling, the assumption of a single client to arrive at each round is often unrealistic. Therefore, as a second research task, we investigate a *batch* variant of the online FLP, adapting both ad-hoc and general purpose algorithms to this case. Accordingly, we evaluate the effect of such a batching technique.

We design four different generators of rewards and costs. Two of them are "adversarial", since they take into account the history and the strategy of the decision maker with the aim of hindering the profit. The other two are stochastic, as they draw rewards and costs according to fixed distributions. We show that for each of the four data generators, a MIP-based algorithm can be chosen that yields better profit than the corresponding ad-hoc ones.

2 Problem Statement and Background

The problem we are interested in is an online version of the well known uncapacitated facility location problem [5] (FLP). In a classical offline maximization setting, we are given a set $I = \{1, \ldots, m\}$ of clients and a set $J = \{1, \ldots, n\}$ of sites which are candidates to host facilities. A vector of rewards $r_i \in \mathbb{R}^n = (r_{i1} \ldots r_{in})$ is given for each client $i \in \{1, \ldots, m\}$ and a vector of opening costs $c \in \mathbb{R}^n = (c_1 \ldots c_n)$ is given for candidate sites. We wish to choose the facilities to open, and assign each client to exactly one open facility, maximizing the total profit, given by the assignment revenues minus the opening costs. In other terms, we want to determine a value for assignment variables x_{ij} and facility variables y_j solving the well-known binary programming model for the FLP. Binary variables y_j are 1 if a facility is opened in j, 0 otherwise. Binary variables x_{ij} are 1 if client i is assigned to an open facility in j, 0 otherwise. Partitioning constraints impose that each client is assigned to exactly one facility, while consistency constraints impose that no client i can be assigned to facility j, unless j is open.

The FLP has been studied also in different online versions, in which clients are assumed to arrive sequentially (see e.g. [2, 6, 7, 10, 18]). The variant we tackle in this paper, which we indicate as OFLP, has been presented in [19] in its basic form and can be formalized as follows. At each trial $t \in \mathbb{N}_+$, the decision maker chooses a set of facilities among n available candidate sites. An energy value $z_j \in [0, 1]$ is associated with each facility $j \in \{1, \dots, n\}$ and the total energy of the open facilities cannot exceed 1. To fix the notation, let y_j^t be 1 if the decision maker has chosen to open facility j and 0 otherwise. After the decision, a client requires to be served. Contextually, for each facility j, two values are revealed: the reward $r_j^t \in [0, 1]$ to assign the client to j and the cost $c_j^t \in [0, 1]$ to open the facility. The client is assigned to the facility with maximum revenue. The decision maker earns a profit given by such a revenue minus the sum of the costs of the facilities which have been opened. The goal of the decision maker is to maximize the overall profit up to trial $T \in \mathbb{N}_+$. We report that such a profit maximization version of FLP is equivalent to the more common minimization version [5] from a modeling point of view, but not from a theoretical guarantees point of view [18].

As reported in [19], in its simplicity, such a setting is able to capture a wide range of practical applications. In order to additionally close the gap between theoretical models and their practical use, we consider its natural generalization, in which more than a single client appears at each trial t. We call it Batch Online Facility Location Problem (BOFLP). Formally, after the decision maker has chosen y_j^t for every j, a fixed number $m \in \mathbb{N}_+$ of clients show up, with revenues $r_{ij}^t \in [0, 1]$ for each $i \in \{1, \dots, m\}$ and $j \in \{1, \dots, n\}$. At the same time, the opening costs $c_j^t \in [0, 1]$ for each $j \in \{1, \dots, n\}$ are revealed. The decision maker gains the profit $\mu^t(y^t) = \sum_{i=1}^m max_{j=1}^n r_{ij}^t y_j^t - \sum_{j=1}^n c_j^t y_j^t$. The definition of cumulative profit is: $P(T) := \sum_{t=1}^T \mu^t(y^t)$.

3 Ad-Hoc OFLP Algorithms

As mentioned in the Introduction, the algorithm MaxHedge [19] is specifically designed for the OFLP. It works by maintaining an internal status, which is represented at each trial t by values w_j^t for each facility j. In the decision phase the interval $U = [0, 1]$ is partitioned in infinitely many sub-intervals U_ℓ for $\ell \in \mathbb{N}_+$, whose length decreases exponentially as ℓ increases. For each sub-interval U_ℓ, some facilities are sampled from the set $\Omega_\ell = \{1 \leq j \leq n \mid z_j \in U_\ell\}$. Only a finite number of sets Ω_ℓ are non-empty. This sampling is performed according to the probability mass function p over Ω_ℓ roughly given by $p(j) = w_j / \sum_{k \in \Omega_\ell} w_k$. For a more detailed description, we refer to the original paper [19]. For every facility j, if j has been sampled at least once, then y_j^t is set to 1 (otherwise it is set to 0). After the decision phase, MaxHedge receives the revenues and the opening costs and executes an update phase of the internal status applying the Projected Online Gradient Descent Algorithm [23].

Theoretical guarantees. MaxHedge ensures theoretical guarantees. It respects a sub-linear pseudo bound on the regret. Formally, let $\mu^t(y) = max_{j=1}^n r_j^t y_j - \sum_{j=1}^n c_j^t y_j$

be the profit function at trial t. It is easy to check that an optimal FLP solution always exists, whose value can be represented as $\mu^t(y)$. In fact, given values for the y_j variables, an optimal solution always exists in which, for each client i, only the x_{ij} variable corresponding to maximum revenue over j is set to 1.

Let F be the set of feasible decisions and let $y_* = max_{y \in F} \sum_{t=1}^{T} \mu^t(y)$ be the best a posteriori solution. A regret bound for the OFLP is of the form:

$$\sum_{t=1}^{T} \mu^t(y_*) - \mathbb{E}\left[\sum_{t=1}^{T} \mu^t(y^t)\right] \leq Cf(T) \tag{1}$$

where C is a constant and f is a function of T. We remark that such a regret bound is proved for y_* (and not, as more commonly, in terms of a generic y). When $f(T)/T$ tends to 0 as T goes to infinity, the regret bound indicates that the algorithm actually "learns" to produce good solutions as the number of trials increases. For MaxHedge, the following slightly different bound holds:

$$\sum_{t=1}^{T} \hat{\mu}^t_{\alpha,\delta}(y) - \mathbb{E}\left[\sum_{t=1}^{T} \mu^t(y^t)\right] \leq n\sqrt{2T}(1-\sqrt{\beta})^2(\hat{r}+\hat{c}) \quad \forall y \in F \tag{2}$$

where $\hat{\mu}^t_{\alpha,\delta}$ are suitable discounted profit functions, $\hat{r} = max_{j \in J, t \leq T}\{r^t_j\}$, $\hat{c} = max_{j \in J, t \leq T}\{c^t_j\}$ and $\beta = max^n_{i=1}\{z_i\}$.

Adapting MaxHedge to the BOFLP. We adapted MaxHedge to solve the BOFLP as follows, obtaining a new class of algorithms we called BatchMaxHedge. Given $k \in \mathbb{N}$ between 1 and m, the decision phase of MaxHedge is executed k consecutive times. For every facility j, if j has been chosen in at least one of the k executions, then y^t_j is set to 1 (and otherwise it is set to 0). The update phase is executed m times sequentially, one for each client i using the corresponding reward values. Every choice of k identifies a different algorithm, that we indicate as BMH(k). Note that when $m = 1$ and $k = 1$, BMH(k) is exactly MaxHedge.

We can give an interpretation of the algorithms corresponding to $k = 1$ and $k = m$. Indeed, the decision process of BMH(1) assumes that the m clients in a batch are not distinguishable one from another. Hence, it makes sense to select the facilities to open like the client was just one. Instead, the decision process of BMH(m) takes into account that, even if the clients cannot be distinguished in advance, every batch of m clients will have a certain variability in its composition. So it is reasonable to select m groups of facilities independently, one for each client, and then merge them.

The worst case complexity of MaxHedge is in $O(nlog(n))$, while for BMH(k) it is in $O(kn + mnlog(n))$, since the decision phase, which takes time $O(n)$ in MaxHedge, in BMH(k) is executed k times, and the update phase is executed m times.

4 MIP-Based Algorithms for BOFLP

Our BOFLP has a natural description in terms of MIP. Indeed, it involves an iterative solution of a sequence of instances of FLP, one for each trial t between 1 and T.

In the literature, a set of well established algorithms based on the iterative solution of mathematical programs is known. The most fundamental algorithm is Follow-The-Leader [11]. In the continuous case with convex profit functions, particularly studied generalizations go under the name of Follow-The-Regularized-Leader [9, 21] or also Dual Averaging [14]. For a comprehensive survey see e.g. [17].

More recently, Follow-The-Leader has been adapted for Combinatorial Optimization. In the worst case scenario, in which there is no assumption about how the feedback to the decision maker is generated, the algorithms are know as Follow-The-Perturbed-Leader (FPL) [15, 16, 22]. In the stochastic feedback setting, algorithms with similar underlying ideas have been proposed [3, 4, 8, 12].

4.1 Follow The Uniformly Perturbed Leader

We considered the class of FPL algorithms introduced in [22], which we called Follow-The-Uniformly-Perturbed-Leader (FUPL). We adapted a representative of that class to solve the BOFLP. Its functioning can be formalized as follows. First of all, define $u_{ij}^1 := 0$ for every $i \in \{1, \ldots, m\}$ and $j \in \{1, \ldots, n\}$ and $v_j^1 := 0$ for every $j \in \{1, \ldots, n\}$. At each trial $t \in \{1, \ldots, T\}$, a decision $y^t \in \{0, 1\}^n$ is made such that, for some $x^t \in \{0, 1\}^{mn}$, (x^t, y^t) is an optimal solution of the following MIP.

$$\max \sum_{i=1}^{m}\sum_{j=1}^{n}\left(u_{ij}^t + U_{ij}(N)\right) x_{ij} - \sum_{j=1}^{n}\left(v_j^t + U_j(N)\right) y_j \qquad \text{(FLPt)}$$

$$\text{s.t.} \sum_{j=1}^{n} x_{ij} = 1 \qquad \forall i \in I \qquad (3)$$

$$x_{ij} \le y_j \qquad \forall i \in I\ \forall j \in J \qquad (4)$$

$$\sum_{j=1}^{n} z_j y_j \le 1 \qquad (5)$$

$$x_{ij} \in \{0, 1\}\ \forall i \in I\ \forall j \in J \qquad y_j \in \{0, 1\}\ \forall j \in J \qquad (6)$$

$U_{ij}(N)$ and $U_j(N)$ indicate real values, sampled independently from random variables with uniform distribution over the set $[0, N]$. The maximum perturbation N is given by

$$N := \left(\frac{4n^2}{eL}\right)^{\frac{1}{3}} T^{\frac{2}{3}}$$

where L is the maximum norm 1 that a feasible solution of FLPt can achieve. At end of trial t, after receiving feedback r^t_{ij} and c^t_j, the decision maker computes $u^t_{ij} = u^{t-1}_{ij} + r^t_{ij}$ and $v^t_j = v^{t-1}_j + c^t_j$.

For Follow-The-Perturbed-Leader algorithms, regret bounds have been proven for linear profit or loss functions. With respect to the setting of the BOFLP, this sentence can be formally expressed as follows. Let us define the linear profit functions

$$\nu^t(x, y) := \sum_{i=1}^{m}\sum_{j=1}^{n} r^t_{ij}x^t_{ij} - \sum_{j=1}^{n} c^t_j y^t_j$$

for every $t \in \{1, \ldots, T\}$, and call C the set of feasible solutions of (6). The cumulative regret is given by:

$$R_\nu(T) = \sum_{t=1}^{T}(\nu^t(x_*, y_*) - \mathbb{E}[\nu^t(x^t, y^t)]), \tag{7}$$

where $(x_*, y_*) \in argmax_{(x,y)\in C}\left[\left(\sum_{t=1}^{T} r^t_{ij}\right) x_{ij} - \left(\sum_{t=1}^{T} c^t_j\right) y_j\right]$. The expected value is intended with respect to the perturbation applied to the learning parameters. It is proved [22] that FUPL guarantees the regret bound:

$$R_\nu(T) \leq 3\left(\frac{L^2n^2}{2e}\right)^{\frac{1}{3}} T^{\frac{2}{3}}. \tag{8}$$

Note that this result does not directly apply to the BOFLP, as in that case the profit function $\mu^t(y)$ is not linear. Anyway, we can observe that ν^t and μ^t are linked: in fact, calculating $\mu^t(y)$ is equivalent to maximize $\nu^t(x, y)$ over C after fixing the variables y. In fact, when $m = 1$, the functions μ^t and ν^t coincide, which means that in this case both FUPL and MaxHedge guarantee some sort of sub-linear regret bound.

We remark that FUPL is an easy algorithm to implement if the solution of (6) is carried out using for example a general purpose MIP solver. However, its worst case time complexity depends on that of (6) itself. In practice, FUPL results efficient only when the concrete instances of (6) that the algorithm encounters are solved efficiently.

4.2 Follow The Clustered Leader

The randomization ingredient of FUPL is essential to ensure the sub-linear regret bound (8) in the worst-case scenario, namely when a hypothetical adversary is able to reproduce every deterministic move of the decision maker. Instead, when the nature is stochastic, the perturbation of the learning parameters is unnecessary, since the generation of the data is blind with respect to the decision making process. FUPL without randomization is a simple Follow-The-Leader (FTL) algorithm. As we show in Sect. 5, FTL is not sufficiently powerful to always provide better profit than BMH(m) when m is greater than 1 and the nature is stochastic. In order to fill the gap, we propose the following variant of FTL, called Follow-The-Clustered-Leader (FCL).

At any iteration t, consider the set of all the reward vectors observed until time step $t-1$, namely $R^t = \{r_i^\tau \mid i \in I,\ 1 \leq \tau \leq t-1\}$. Assume that every single reward value is inversely proportional to the distance between the relative client and facility. Then it is reasonable to cluster the reward vectors so that different clusters identify different regions in the space of the facilities, and therefore different typologies of clients.

Given the set $H^t := \{1, \ldots, \ell^t\}$, we apply the K-means algorithm to obtain the clusters $C_1^t, \ldots, C_{\ell^t}^t$. Let assume none of them to be empty. Then we compute the surrogate rewards $\bar{r}_{hj}^t := \dfrac{1}{t-1} \sum_{r \in C_h^t} r_j$ for every $h \in H^t$ and $j \in J$. We also compute the average opening costs $\bar{c}_j^t := \dfrac{1}{t-1} \sum_{\tau<t} c_j^\tau$. In this way we can solve the following model.

$$\max \sum_{h=1}^{\ell^t} \sum_{j=1}^{n} \bar{r}_{hj}^t x_{hj} - \sum_{j=1}^{n} \bar{c}_j^t y_j \qquad \text{(FLPt')}$$

$$s.t. (5') \ (6') \ (7') \ (8')$$

where (5'), (6'), (7'), (8') are the analogous constraints of (5), (6), (7), (8) with H^t in place of I.

Since the number of clusters to use is unknown, we define a constant $\Delta \in \mathbb{N}_+$ and increase ℓ^t by 1 every Δ iterations according to $\ell^t := \lfloor t/\Delta \rfloor + 1$. Moreover, since the clustering operation is expensive, we update the clusters and the surrogate rewards less often as the time step increases, relaying on the fact that the quantity of reward vectors collected increases as well and therefore additional vectors will have in general less impact on the estimation quality that we can obtain. In particular, we compute the clusters only when $t = 1$ (trivial case) or when

$$t \equiv 0 \ (mod \ m(\lfloor t/\Delta \rfloor + 1)).$$

5 Experimental Setting and Results

We remark that no experimental performance evaluation of MaxHedge has been presented in [19] nor in any other work, to the best of our knowledge. Similarly, no specific experiments on MIP-based algorithms for the OFLP have been presented so far. Our experimental analysis has therefore two targets: (a) the assessment of the relative performance of these two classes of algorithms and (b) the evaluation of the effect of moving from a OFLP to a BOFLP model. Our experiments have been conducted on a machine with CPU AMD Ryzen 7 5800H and O.S. Ubuntu 23.04. The binary linear programming models have been solved using Gurobi version 9.5.1 and the code has been written in Python 3.10.

5.1 Experimental Test Bed

Three of our test cases (AAH, AAF, MPSN - see below) share the same set of facility locations. We choose $n = 20$ and for each $j \in J$ we extracted a pair $\zeta^{(j)} = (\zeta_1^{(j)}, \zeta_2^{(j)})$ from $[0, 1]^2$ with uniform distribution and we fixed them once for all. A random generator consisting of a Python re-implementation of GEN2 [20], was used to generate a knapsack instance (v, w, c) where v are the values of the items, w the weights and c is the capacity. Then we normalized the vectors dividing v by the maximum possible item value and w by c. Finally, we fixed the mean opening cost vector $\bar{c} := v$ and the energy vector $z := w$.

We designed four different online rewards-and-costs generators (so called *natures*), in part inspired by the experimental setting presented in [22]. Two of these generators are "adversarial" and two of them are stochastic. One of the stochastic generators is based on real data [1].

Adversary Against History (AAH). The first adversarial nature tries to damage the decision maker taking into account the whole history of the decisions under the assumption that the next decision will not differ too much from the previous ones. For each $i \in I$ this nature considers $x_i = \sum_{\tau=1}^{t-1} x_i^\tau$ and its maximum value $X_i = max_{i=1}^n x_i$ and then it estimates the next decision that will be taken by the decision maker. To do so, it produces a vector $\hat{y}_i^t \in \{0, 1\}^n$ such that $\hat{y}_{ij}^t = 1$ with probability x_{ij}/X_i and 0 otherwise. At this point, the nature finds the farthest point γ_i^t in $[0, 1]^2$ from all the open facilities according to $\hat{y}_i^t$. It assigns $r_{ij}^t := \sqrt{2} - \|\gamma_i^t - \zeta^{(j)}\|_2$ for all $j \in J$. At the end, the nature generates a further decision estimate $\hat{y}^t$ and for each $j \in J$ assigns $c_j^t := 1$ if $\hat{y}_j^t = 1$ and 0 otherwise.

Adversary Against the Future (AAF). This adversarial nature acts exactly like AAH except that the vectors of open facilities, $\hat{y}_i^t$ for $i \in I$ and $\hat{y}^t$, are obtained mimicking the method the decision maker would use to calculate y^t. In other words it applies the same decision making step of the algorithm used by the decision maker. Note that if the algorithm is FUPL, the nature cannot perturb the learning parameters u^t

and v^t using the same random values of the decision maker, because it is impossible to know in advance what such values will be. So the nature has to adopt its own random values.

Multi-Pyramidal Stochastic Nature (MPSN). This stochastic nature acts in the following way. Three fixed available facilities with coordinates $\zeta^{(\tilde{j}_1)}$, $\zeta^{(\tilde{j}_2)}$ and $\zeta^{(\tilde{j}_3)}$ are considered. Each of them has an attractiveness, respectively $\alpha^{(\tilde{j}_1)}$, $\alpha^{(\tilde{j}_2)}$ and $\alpha^{(\tilde{j}_3)}$, such that $\alpha^{(\tilde{j}_1)} + \alpha^{(\tilde{j}_2)} + \alpha^{(\tilde{j}_3)} = 1$. For each $i \in I$ the nature samples an index $\tilde{j}^t$ from $\{\tilde{j}_1, \tilde{j}_2, \tilde{j}_3\}$ according to the probability distribution given by the respective attractiveness values. It extracts a point $\gamma_i^t = (\gamma_{i1}^t, \gamma_{i2}^t)$ from $[0, 1]^2$ by sampling γ_{i1}^t from the triangular distribution on [0, 1] with mode $\zeta_1^{(\tilde{j}^t)}$ and γ_{i2}^t from the triangular distribution on [0, 1] with mode $\zeta_2^{(\tilde{j}^t)}$. At last, it assigns $r_{ij}^t := \sqrt{2} - \|\gamma_i^t - \zeta^{(j)}\|_2$ for each $j \in J$. Regarding the opening costs, the nature randomly defines $c_j^t := max(0, min(1, \bar{c}_j + U_{[-0.3,0.3]}))$ for all $j \in J$, where $U_{[-0.3,0.3]}$ indicates a real value sampled uniformly from $[-0.3, 0.3]$.

Telecom Stochastic Nature (TSN). The second stochastic nature is based on real data, and can be sketched as follows. The data comes from the Telecom Italia Big Data Challange Dataset [1]. In particular, we used the information regarding the number of Call Detail Records (CDRs) for internet connections registered to the mobile network of the telecommunication company Telecom Italia in the city of Milan (Italy) from November 1st 2013 to January 1st 2014. Both space and time was discretized in square cells of 235 m approximately, and time slots of 10 min, respectively. We placed 20 candidate location sites in a regular fashion. Data was manipulated to obtain probability distributions, which changes every 60 iterations simulating the passage of time from a 10 min interval to the next one. Positions of clients are sampled according to these distributions. Then the rewards as well as the opening costs are generated in the same way as done by MPSN.

5.2 *Evaluation of the Algorithms*

In the single client case, we compared BMH(1) (MaxHedge) with FTPL in the adversarial setting and with FTL in the stochastic one. Then we applied BMH(1), BMH(m), FUPL and FCL for $m \in \{10, 20, 40\}$, both in the adversarial and the stochastic setting. For FCL, the value of the parameter Δ has been set to 1000. All the algorithms have been applied with respect to the same profit function $\mu^t(y)$ as defined in Sect. 3 and have been evaluated on the profit obtained as function of the iteration counter and on the computing time. In Figs. 1, 2, 3 and 4 we report a selection of our results. Each line corresponds to an algorithm: BMH(1) (orange), BMH(m) (red), FUPL (black), FTL (blue), FCL (green). Horizontal axes represent time iterations from 1 to 10.000, vertical axes represent profit (Figs. 1, 2, 3 and 4). Each figure refers to specific combinations of batch sizes and natures, as indicated in the related caption. All the time series have been averaged with a sliding window of 100 iterations.

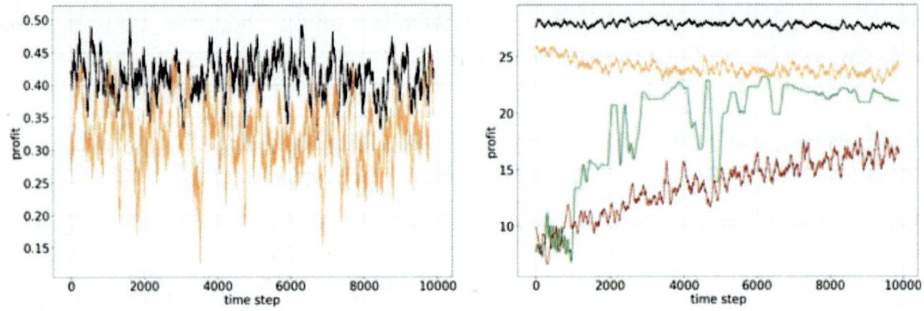

Fig. 1 Profit per iteration, AAF: $m = 1$ (left) and $m = 40$ (right)

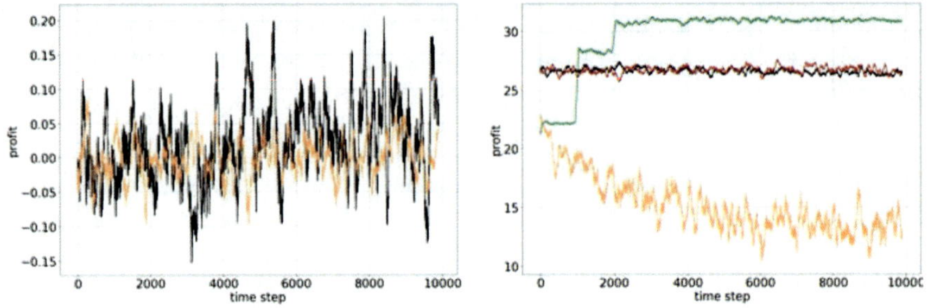

Fig. 2 Profit per iteration, AAH: $m = 1$ (left) and $m = 40$ (right)

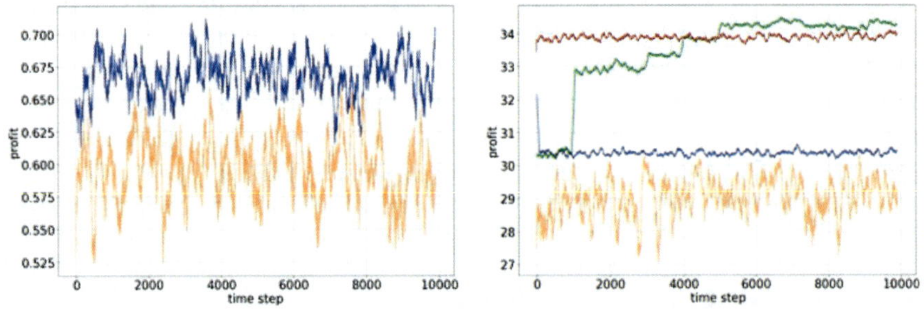

Fig. 3 Profit per iteration, MPSN: $m = 1$ (left) and $m = 40$ (right)

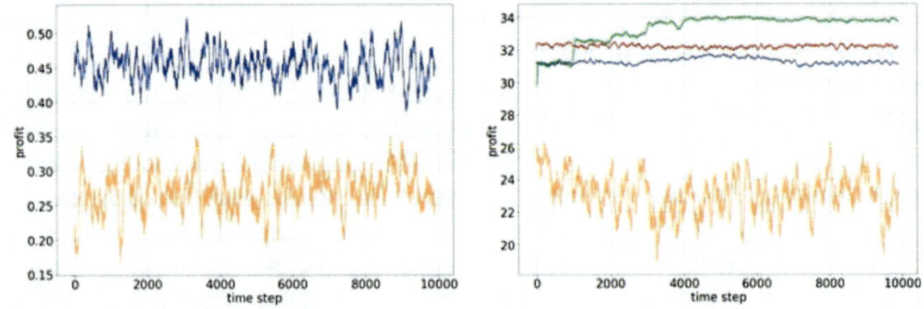

Fig. 4 Profit per iteration, TSN: $m = 1$ (left) and $m = 40$ (right)

Profit. On the long run, FUPL obtains higher profit than BMH(1) in the adversarial setting with single client, and similarly FTL does in the stochastic case. For every choice of $m \in \{10, 20, 40\}$, FUPL is the best of the four tested algorithms when the nature is AAF. For the other natures, FCL obtains the best performance starting from when the number of clusters is sufficiently high. We observed that such a number is directly proportional to the number of clients per iteration. We also noticed that when the nature is AAH, FUPL is more profitable than the two BMH representatives for $m \in \{10, 20\}$, but its performance looks equivalent to that of BMH(m) for $m = 40$. We calculated the average percentage profit error with respect to the best maximum obtainable profit. In general it is higher with adversarial generators and more restrained with stochastic ones. Finally, we found that grouping clients into batches is beneficial, since it increases the profit per client and decreases the average percent error per client.

Running time. We report that the algorithms required fractions of seconds per iteration on average to run, except for FUPL with $m = 40$. When $m = 1$, BMH(1) was faster while FUPL required less than 7 ms per and FTL less than 4 ms. When $m = 40$ and the nature is AAF, FUPL took between 1.4 and 1.6 seconds per iteration, while the other algorithms stayed under 0.6 s. When the nature is AAH, all the four algorithms required less than 0.2 s. With MPSN, they took no more than 70 ms. With TSN, they have not exceed 90 ms. The running time of FCL approximately grows during all the execution, which reflects our choice to gradually increase the number of clusters.

6 Conclusions

The paper focused on Online algorithms for Facility Location Problems, and more specifically on two research questions.

The first one was to understand the experimental behaviour of both ad-hoc and MIP-based algorithms. We designed four test scenarios, by considering two adversarial natures, and two stochastic natures, one of which coming from real data. We found out that for each case there is a MIP-based approach that provides better profit. The CPU time required by the MIP-based algorithms is comparable to that of ad-hoc algorithms, with the exception of the adversary against the future.

The second research question was to adapt the existing methods to handle batches of clients at each decision round (instead of one, as assumed in the literature). We indeed found adaptations to be possible without changing too much the nature of the algorithms. Furthermore, we found such a batching technique useful to obtain higher profit per client with lower average percent profit error.

References

1. Barlacchi, G., De Nadai, M., Larcher, R., Casella, A., Chitic, C., Torrisi, G., Antonelli, F., Vespignani, A., 'Sandy' Pentland, A., Lepri, B.: A multi-source dataset of urban life in the city of Milan and the Province of Trentino. Sci. Data **2** (2015)
2. Chakraborty, A., Vaze, R.: Online facility location with timed-requests and congestion (2022). arXiv:2211.11961
3. Chen, W., Wang, Y., Yuan, Y.: Combinatorial multi-armed bandit: general framework, results and applications. In: International Conference on Machine Learning (2013)
4. Combes, R., Talebi, M.S., Proutière, A., Lelarge, M.: Combinatorial bandits revisited. In: Conference on Neural Information Processing Systems (2015)
5. Drezner, Z., Hamacher, H.W.: Facility location applications and theory. Springer, Berlin, Heidelberg (2002)
6. Feldkord, B., Meyer auf der Heide, F.: Online facility location with mobile facilities. In: Proceedings of the 30th on Symposium on Parallelism in Algorithms and Architectures (2018)
7. Fotakis, D., Kavouras, L., Zakynthinou, L.: Online facility location in evolving metrics. Algorithms **14**, 73 (2021)
8. Gai, Y., Krishnamachari, B., Jain, R.: Combinatorial network optimization with unknown variables: multi-armed bandits with linear rewards and individual observations. IEEE/ACM Trans. Netw. **20**, 1466–1478 (2012)
9. Hazan, E., Kale, S.: Extracting certainty from uncertainty: regret bounded by variation in costs. Mach. Learn. **80**, 165–188 (2008)
10. Jiang, S.H.-C., Liu, E., Lyu, Y., Tang, Z.G., Zhang, Y.: Online facility location with predictions (2021). arXiv:2110.08840
11. Kalai, A.T., Vempala, S.S.: Efficient algorithms for online decision problems. J. Comput. Syst. Sci. **71**, 291–307 (2005)
12. Kveton, B., Wen, Z., Ashkan, A., Szepesvari, C.: Tight regret bounds for stochastic combinatorial semi-bandits. In: International Conference on Artificial Intelligence and Statistics (2015)
13. Meyerson, A.: Online facility location. In: Proceedings 42nd IEEE Symposium on Foundations of Computer Science, pp. 426–431 (2001)
14. Nesterov, Y.: Primal-dual subgradient methods for convex problems. Math. Program. **120**, 221–259 (2005)
15. Neu, G., Bartók, G.: Importance weighting without importance weights: an efficient algorithm for combinatorial semi-bandits. J. Mach. Learn. Res. **17**, 1–21 (2016)
16. Neu, G., Valko, M.: Online combinatorial optimization with stochastic decision sets and adversarial losses. In: NIPS (2014)
17. Orabona, F.: A modern introduction to online learning (2022). arXiv:1912.13213
18. Pasteris, S., He, T., Vitale, F., Wang, S., Herbster, M.: Online learning of facility locations. In: International Conference on Algorithmic Learning Theory (2021)
19. Pasteris, S., Vitale, F., Chan, K.S., Wang, S., Herbster, M.: Maxhedge: maximising a maximum online. In: International Conference on Artificial Intelligence and Statistics (2019)
20. Pisinger, D.: David Pisinger's optimization codes (1994)
21. Shalev-Shwartz, S., Singer, Y.: A primal-dual perspective of online learning algorithms. Mach. Learn. **69**, 115–142 (2007)
22. Yang, F., Chen, W., Zhang, J., Sun, X.: Follow the perturbed approximate leader for solving semi-bandit combinatorial optimization. Front. Comput. Sci. **15**, 155404 (2021)
23. Zinkevich, M.A.: Online convex programming and generalized infinitesimal gradient ascent. In: International Conference on Machine Learning (2003)

Data-Driven Feasibility for the Resource Constrained Shortest Path Problem

Cristina Ondei, Alberto Ceselli, and Marco Trubian

Abstract Resource Constrained Shortest Path Problems (RCSPP) have wide applicability, representing a flexible model for network applications. Furthermore, they frequently arise as subproblems in decomposition-based methods, as occurs in column generation for Vehicle Routing Problems. In all these settings, being able to perform early detection of infeasibility helps to strongly reduce computing times. For instance, dynamic programming is often used to design RCSPP algorithms: labels representing partial solutions are iteratively created and extended, and these can be dropped if they are found to have no feasible (and profitable) completion. Many fathoming heuristics have been proposed in the literature. We experiment a data-driven approach in this context, using supervised learning models to deal with the problem of detecting infeasibility. We design features which are not dependent on instance size, having different computing cost. We compare the tradeoff between computational effort and performance which can be achieved, when a binary classifier is employed. Our results indicate such an attempt to be effective.

1 Introduction

In the Resource Constrained Shortest Path Problem (RCSPP) [1] a graph is given, having two special source and sink nodes. Furthermore, a set of *resources* is given: a vector of values is associated to each edge, having one element for each resource. A consumption limit for each resource is finally given. A path is considered to be *feasible* if the sum of resource values on its edges exceeds the corresponding limit

C. Ondei (✉) · A. Ceselli · M. Trubian
Department of Computer Science,University of Milan, Milan, Italy
e-mail: cristina.ondei@unimi.it

A. Ceselli
e-mail: alberto.ceselli@unimi.it

M. Trubian
e-mail: marco.trubian@unimi.it

A. Brieden et al. (eds.), *Graphs and Combinatorial Optimization: from Theory to Applications*, AIRO Springer Series 13,
https://doi.org/10.1007/978-3-031-46826-1_11

for no resource. In its feasibility version, the RCSPP asks whether a feasible path from source to sink exists or not. In its optimization version, one resource is given interpretation as a *cost*: the aim is to find a feasible path from source to sink whose sum of cost values on edges is minimized.

Besides its direct mapping to a routing problem on graphs, the RCSPP has applications as other standalone combinatorial optimization problems, for instance in the realm of project scheduling. It is more famous, however, as a subproblem of Vehicle Routing Problems [2] when decomposition algorithms (such as column generation ones) are employed [3]. In these algorithms, RCSPPs are in fact iteratively solved as pricing subproblems, when feasible routes (e.g. respecting overall time and capacity limits) need to be generated, having negative reduced cost.

The RCSPP is weakly NP-Hard [1]; dynamic programming is often used for its optimal resolution. A well known technique for speeding up these algorithms is the so-called *completion bounding* [4]: for each partial path a check is performed, to understand if any extension may exist leading to a feasible complete path. These techniques may become key ingredients when RCSPPs need to be solved iteratively, as in column generation algorithms. Since this bounding check may be performed a very high number of times (e.g. at each extension operation of a dynamic programming algorithm) its speed is crucial. Machine learning approaches have also been explored in this context. For instance, in [5] binary classifiers are used to choose the critical resource in a bidirectional label setting algorithm; in [6] machine learning is used to select the most promising arcs when RCSPP is the pricing subproblem in a column generation algorithm.

We propose to tackle the RCSPP feasibility problem with data-driven models, and in particular with supervised learning ones [7]. In Sect. 2 we formally describe the problem. In Sect. 3 we introduce a set of *features*, which become parameters that we observe to predict feasibility or infeasibility of a path. In Sect. 4 we report on a set of experiments where *binary classifiers* are trained by sampling both feasible and infeasible RCSPP solutions and measuring their features; these classifiers are then tested on new RCSPP instances to understand their feasibility predictive power. In Sect. 5 we collect some brief conclusions.

2 Problem Definition

Let $\mathcal{G} = (\mathcal{V}, \mathcal{E})$ be an undirected weighted graph, with $n = |\mathcal{V}|$ and $m = |\mathcal{E}|$, and let $\mathcal{H}$ be a set of r resources. A positive cost c_e is associated with each edge $e \in \mathcal{E}$. For every resource $h \in \mathcal{H}$ and for every edge $e \in \mathcal{E}$ we consider a positive weight w_e^h, representing the consumption of resource h traversing edge e. We assume a consumption limit W^h for every resource $h \in \mathcal{H}$. Let the vertices $s \in \mathcal{V}$ and $t \in \mathcal{V}$ be respectively the source and the destination.

A feasible solution is a path $\mathcal{P}^{st}$ between vertex s and vertex t that complies with the capacity constraint on every resource, i.e. $\sum_{e \in \mathcal{P}^{st}} w_e^h \leq W^h$ for each $h \in \mathcal{H}$.

The optimization problem traditionally consists in finding the minimum cost path. In our framework, the focus is on the feasibility problem: given an instance for the RCSPP, determine whether it admits a feasible solution or not.

It has been proven that RCSPP, in its feasibility formulation, is an NP-complete problem, even in the case with only two resources [1].

Let i be an arbitrary instance of the RCSPP. We consider a vector of *features* $\mathbf{x}_i \in \mathbb{R}^p$. Each of these features measures a particular parameter of the instance i. We assume that a function $g : \mathbb{R}^p \rightarrow \{F, I\}$ exists. The function g maps each feature vector $\mathbf{x}_i$ to a label F if the instance i admits a feasible solution, I if it does not.

Our objective is to use supervised learning algorithms to find (an approximation of) such a function $g()$. In particular, we exploit supervised learning: we collect a set of RCSPP instances $\mathcal{I}$ for which feasibility *is known*. We use such a *training* set to build an approximation of $g()$. Then, we exploit $g()$ to actually perform feasibility predictions on new RCSPP instances.

The most critical point of this approach is the so-called *feature engineering* process, that is the choice of parameters to measure on RCSPP instances, which is discussed in Sect. 3. We remark that, hypothetically speaking, it would be easy to build a perfect function $g()$ by including in these features the *actual exact resolution* of the RCSPP. Indeed, these features must balance predictive power with computational efficiency: our aim is to obtain high predictive power with a small fraction of the computing effort needed to actually solve the RCSPP.

3 Feature Engineering

For our approach to be effective, we designed features that are quickly computable and not dependent on specific aspects of the instance, such as the dimension of the graph and the range of the parameters. This allows the approach to be *scalable*, and potentially to be *generalized* to instances which are *very different* from those in the training set.

Most of the features considered are based on single resource shortest path computations. For each resource $k \in \mathcal{H}$, we indicate with $\mathcal{P}_k^{st}$ the shortest path between s and t, with respect to the resource $k \in \mathcal{H}$. The Shortest Path Problem is known to be solvable in polynomial time; moreover, several advanced algorithmic techniques make it very fast also empirically. We finally mention that in column generation contexts, as those discussed in the introduction, single resource shortest paths for each resource besides costs can be computed once in a preprocessing phase.

To effectively predict feasibility, we need to exploit the aspects of the instance that could indicate the existence of a feasible solution. In this context, the single resource paths could give useful information. With our features we tried to summarize important aspects, such as the structure and the resource consumption along these paths. Our features are based on the idea that if a feasible solution exits, it might be close to these paths.

We indicate with u_k^h the consumption of resource $h \in \mathcal{H}$ along $\mathcal{P}_k^{st}$ and with $\bar{u}_k^h$ the average consumption of resource $h \in \mathcal{H}$ on each edge of the path $\mathcal{P}_k^{st}$. We indicate with l_k the length (i.e. the number of edges) of the path and with $\mathcal{V}_k$ the set of its vertices.

Finally, let d be the diameter of the graph and $\bar{w}^h$ the average consumption of resource $h \in \mathcal{H}$ on each edge of the graph.

With this notation, we can formally define the features we designed. We grouped them into three classes.

Path related features.

The first group of features are strictly related to the shortest paths and the resource consumption along these paths.

Feature 1 Relative consumption of resource h traversing the path $\mathcal{P}_k^{st}$.

$$F_{hk}^1 = \frac{u_k^h}{W^h} \qquad \forall h \in \mathcal{H}, \forall k \in \mathcal{H} \tag{1}$$

Observation 1 If a resource $k \in \mathcal{H}$ exists, such that $F_{hk}^1 \leq 1 \; \forall h \in \mathcal{H}$, then the path $\mathcal{P}_k^{st}$ is a feasible solution for the RCSPP.

Observation 2 If a resource $k \in \mathcal{H}$ exists, such that $F_{kk}^2 > 1$, then the instance is infeasible.

We remark that these two observations span only specific cases. In particular, Observation 1 is a sufficient, but not necessary, condition for feasibility. In the general one, a feasible RCSPP solution may exist, crossing parts of $\mathcal{P}_k^{st}$ for different k.

Still, this features give a measure of how much of each resource is consumed.

Since these features can exactly establish feasibility, as shown in Observations 1 and 2, we expect them to be highly predictive.

Feature 2 Average relative consumption of resource h traversing the path $\mathcal{P}_k^{st}$.

$$F_{hk}^2 = \frac{\bar{u}_k^h}{W^h} \qquad \forall h \in \mathcal{H}, \forall k \in \mathcal{H} \tag{2}$$

Path-Graph related features.

In the second group we have features that consider both path related and graph related aspects. They are designed to exploit some topological aspects of the graph, concerning the vertices and edges in the single-resource optimal paths.

Definition 1 Given a resource $h \in \mathcal{H}$, for each edge $e = (i, j) \in \mathcal{E}$, we define *reduced cost of edge e with respect to resource h* the quantity $rc^h(i, j) = w_e^h - (w^h(\mathcal{P}^{sj}) - w^h(\mathcal{P}^{si}))$, where $w^h(\mathcal{P}^{si})$ and $w^h(\mathcal{P}^{sj})$ represent the consumption of resource h along the minimum paths, w.r.t resource h, between s and i and j respectively.

If the edge $e = (i, j) \in \mathcal{P}_h^{st}$, then the reduced cost $rc^h(i, j) = 0$. Otherwise, it is $rc^h(i, j) > 0$ when the edge (i, j) does not belong to the shortest path between s and j w.r.t. resource $h \in \mathcal{H}$. In this case the reduced cost represents the difference between the value of the path between s and j passing through edge (i, j) and the value of the shortest path between s and j.

We consider the quantity rc_i^h for each $i \in \mathcal{V}_h$, as the minimum reduced cost of the outgoing edges of i not in $\mathcal{P}_h^{st}$. We indicate with rc^h the average, on the vertices of the path $\mathcal{P}_h^{st}$, of the quantities rc_i^h.

Definition 2 Given two vertices $i, j \in \mathcal{V}$, we define the *distance* $d(i, j)$ between i and j as the minimum number of edges in a path between i and j.

Definition 3 We define *D-neighbourhood* of $\mathcal{P}_h^{st}$ the subgraph induced by the vertices that have distance from the vertices in $\mathcal{P}_h^{st}$ less or equal than D (w.r.t. the distance in Definition 2).

Feature 3 Average reduced cost on the vertices of path $\mathcal{P}_h^{st}$, considering for each vertex the outgoing edge not in $\mathcal{P}_h^{st}$ with minimum reduced cost.

$$F_h^3 = \frac{rc^h}{\bar{w}^h} \qquad h \in \mathcal{H} \tag{3}$$

The reduced cost for an edge $e = (i, j)$ represents a penalty paid by deviating from the shortest path between s and j. Lower values of this feature could indicate the presence of non-expensive deviations from the path $\mathcal{P}_h^{st}$ and the presence of a feasible solution.

The value is normalized, in order to reduce the dependency of this feature from the range of the resource values.

Feature 4 Ratio between the average consumption of resource $h \in \mathcal{H}$ along path $\mathcal{P}_h^{st}$ and the average consumption of resource $h \in \mathcal{H}$ along the edges in a D-neighbourhood of $\mathcal{P}_h^{st}$, not in the path itself.

$$F_{hD}^4 = \frac{\bar{u}_h^h}{\bar{u}_D^h} \qquad h \in \mathcal{H} \tag{4}$$

This feature measures the difference between the resource consumption on the edges in the path and the consumption on the unused edges. The idea is that if the edges in the path are much more convenient than the others, it would be harder to find a feasible solution.

It is worth noticing that the distance in Definition 2 is well defined only in undirected graphs, otherwise the symmetry property would not hold. In case of an application on directed graphs, this feature should be redesigned using a different distance.

In our analysis we considered neighbourhoods with $D = 2, 3$.

Feature 5 Percentage of edges used in path $\mathcal{P}_h^{st}$

$$F_h^5 = \frac{l_h}{m} \qquad \forall h \in \mathcal{H} \tag{5}$$

This feature lies on the assumption that the more edges are used in the shortest paths, less likely there would be different feasible paths.

Feature 6 Maximum, minimum and average degree of the vertices in $\mathcal{P}_h^{st}$

$$F^6_{h,max} = \max_{i \in \mathcal{V}_h} deg(i) \qquad \forall h \in \mathcal{H} \tag{6a}$$

$$F^6_{h,min} = \min_{i \in \mathcal{V}_h} deg(i) \qquad \forall h \in \mathcal{H} \tag{6b}$$

$$F^6_{h,avg} = \frac{\sum_{i \in \mathcal{V}_h} deg(i)}{l_h + 1} \qquad \forall h \in \mathcal{H} \tag{6c}$$

The main idea behind this feature is that if the vertices in the paths have many connections, it would be easier to find a feasible solution.

Graph related features.

The last group of features considers only some topological aspects of the graph.

Feature 7 Comparison of the distance between s and t and the diameter of the graph.

$$F^7 = \frac{d(s,t)}{d} \tag{7}$$

This feature represents a measure of the distance of the source s and the destination t in the graph. We expect higher values to be linked to infeasibility: if s and t are far away in the graph it would be harder to find a feasible solution.

Feature 8 Maximum, minimum and average degree of the vertices in N

$$F^8_{max} = \max_{i \in N} deg(i) \tag{8a}$$

$$F^8_{min} = \min_{i \in N} deg(i) \tag{8b}$$

$$F^8_{avg} = \frac{\sum_{i \in N} deg(i)}{n} \tag{8c}$$

These features give a measure of the sparsity of the graph. Similarly to features F^6_{max}, F^6_{min} and F^6_{avg}, the idea is that more connections could facilitate the presence of a feasible solution.

4 Computational Experiments

Dataset description.

For the experimental analysis we considered the RCSPP with two resources. To train and test the machine learning algorithms, we build a dataset of 9600 instances, that have been correctly classified as *feasible* or *infeasible*.

The graphs for the instances were extracted as random subgraphs of the instances for the *9th DIMACS Implementation Challenge* [8]. We created 20 graphs of 5500 vertices each. Starting from the same base graph, we generated various instances, by assigning different weights to the edges and selecting different sources, destinations and resource limits. In particular, our dataset is composed by different scenarios of increasing challenge:

- instances that differ only in the resource limits;
- instances with different source, destination and resource limits;
- instances based on the same graph with different resource values on the edges;
- instances based on entirely different graphs (with different resource values on the edges).

We established the feasibility of each instance by solving the corresponding optimization problem, considering one of the resources as the cost to be minimized. A mixed integer program for the RCSPP was implemented in Python, using the Pyomo library [9, 10] e, and solved with CPLEX 20.1.

The instances resulted fairly balanced between feasible (59.6%) and infeasible (40.4%).

Features computation.

In Table 1 we summarize the average computing time and the worst case complexity for the computation of the features.

Considering different instances are based on the same graph, we could compute some static graph indexes in a preprocessing phase. This is often the case when the RCSPP is solved as a subproblem. In particular, for the worst case complexities reported in Table 1, we assume the following quantities to be precomputed:

- Diameter of the graph;
- Average resource consumption on the edges $\bar{u}_k^h$, for each $h, k \in \mathcal{H}$;
- Minimum, maximum and average degree of the vertices of the graph.

In the first line of Table 1 we report the theoretical complexity and the average computation times for the shortest paths. They are solved using Dijkstra-like algorithms, for which we did not invest in sophisticated implementations.

We assume that the shortest paths are computed once for every instance. Thus their computation time is not included in the remaining lines of Table 1. We report both the complexity required to compute the single feature and the overall complexity,

Table 1 Worst case complexity and average computing time of the shortest paths and the features

Features	Worst case complexity (single feature)	Average computing time (s)	Total worst case complexity
Shortest path $\mathcal{P}_h^{st}$ $h \in \mathcal{H}$	$O(m \log n)$	$1.333 \cdot 10^{-2}$	$O(rm \log n)$
F_{hh}^1 $h \in \mathcal{H}$	$O(1)$	$1.658 \cdot 10^{-6}$	$O(r)$
F_{hk}^1 $h, k \in \mathcal{H}, h \neq k$	$O(n)$	$1.398 \cdot 10^{-5}$	$O(r^2 n)$
F_{hk}^2 $h, k \in \mathcal{H}$	$O(1)$	$5.283 \cdot 10^{-7}$	$O(r^2)$
F_h^3 $h \in \mathcal{H}$	$O(m)$	$7.243 \cdot 10^{-5}$	$O(rm)$
F_{hD}^4 $h \in \mathcal{H}, D = 2, 3$	$O(n + m)$	$1.087 \cdot 10^{-3}$	$O((n + m)r\|D\|)$
F_h^5 $h \in \mathcal{H}$	$O(1)$	$6.776 \cdot 10^{-7}$	$O(r)$
$F_{h,max}^6$ $F_{h,min}^6$ $F_{h,avg}^6$ $h \in \mathcal{H}$	$O(m)$	$8.306 \cdot 10^{-5}$	$O(rm)$
F^7	$O(m + n)$	$2.327 \cdot 10^{-3}$	$O(m + n)$
F_{max}^8 F_{min}^8 F_{avg}^8	$O(1)$	$2.551 \cdot 10^{-7}$	$O(1)$

considering many features are computed for each resource. The total computing time for each instance is about 0.064 s, on average.

Prediction phase.

We tested our features on three models: a Decision Tree with different impurity functions, Support Vector Machines with different kernel functions, and a Gradient Boosting (Random Forest) Classifier. The models were implemented in Python using Scikit-Learn library [11].

To enforce the results, we implemented a Cross Validation procedure. The dataset was randomly split into five groups of the same size. At each iteration, four groups were used for the training of the model, the remaining group for the test. The scores reported in Tables 2 and 3 are the average ones, obtained at each iteration.

For the first run of experiments, we randomly split the dataset between training and testing. Results are shown in Table 2. The best performing model is the Gradient Boosting, reaching an accuracy of 84% on the test set.

In the second round of experiments, we split the dataset in order to have instances from different groups based on different graphs. This leads to test the model on instances that are not based on the same graphs as the instances used for the training. In Table 3 are summarized the results. The scores are comparable to the previous case, showing that the efficacy of our features does not depend on having the same underlying graph. The best performing model is again the Gradient Boosting.

Features importance analysis.

We performed an analysis of the importance of the features, to see if they are all relevant in the effectiveness of the predictions. We focused on the Gradient Boosting model, that was the best performing one.

Table 2 Cross validated scores

Model		Accuracy train (%)	Accuracy test (%)	I_precision (%)	F_precision (%)	I_recall (%)	F_recall (%)	F1-score (%)
Decision Tree	Gini	81.47	80.01	75.66	82.98	74.59	83.73	75.05
	Entropy	81.07	79.77	74.64	83.41	75.77	82.47	75.15
Support Vector Machines	Linear	79.75	79.60	74.55	83.07	75.10	82.67	74.81
	Poly 2	80.81	80.20	75.72	83.20	75.02	83.74	75.34
	Poly 3	81.09	80.47	76.06	83.42	73.35	83.96	75.68
	rbf	81.02	80.25	75.70	83.30	75.23	83.68	75.44
Gradient boosting		88.24	84.23	81.57	85.94	78.75	87.96	80.11

Table 3 Cross validated scores. In this case training instances and test instances are based on different graphs

Model		Accuracy train (%)	Accuracy Test (%)	I_precision (%)	F_precision (%)	I_recall (%)	F_recall (%)	F1-score (%)
Decision Tree	Gini	80.86	79.54	74.07	83.34	75.76	82.02	74.91
	Entropy	80.59	79.43	74.55	82.84	74.68	82.51	74.55
Support Vector Machines	Linear	79.77	79.58	74.66	82.93	74.87	82.70	74.74
	Poly 2	80.73	79.81	75.61	82.57	73.86	83.77	74.69
	Poly 3	81.05	79.72	75.70	82.28	73.25	84.01	74.44
	rbf	81.13	79.67	75.42	82.51	73.75	83.61	74.51
Gradient boosting		88.41	84.10	81.53	85.72	78.29	88.00	79.86

We performed a Recursive Feature Elimination using a Gradient Boosting model: starting with all the features, at each iteration the model is trained and the least important feature is discarded. In Table 4 we rank the features by decreasing importance: the first one was the last remaining feature in the recursive elimination. In Table 4 we also report the average importance of the features in the full Gradient Boosting model, i.e. the model trained with all the features. The results are coherent with the rank obtained by Recursive Feature Elimination.

In Fig. 1 we report the accuracy (on the test set) of the model w.r.t. the number of features retained. We see that 5 features are necessary to reach a high accuracy level. There is an improvement up until 10 features. Adding further features seems irrelevant.

In Table 5 we report the results for the Gradient Boosting using only the best 10 features (considering the rank in Table 4). The scores are comparable to those obtained in the full model, reported in Tables 2 and 3.

Finally, considering the model with the selected features, we tested the impact of these features, by removing one feature at a time. In Fig. 2 we represent the accuracy gap on the test set, meaning the difference between the test accuracy of model trained with the 10 selected features and the test accuracy of the model trained

Table 4 Rank of the features obtained by Recursive Feature Elimination and average importance in the full model. The last 5 features are not reported, since they have an importance below 0.001

Feature	Rank in recursive feature elimination	Importance
F_{11}^1	1	0,391
F_{22}^1	2	0,254
F_{21}^1	3	0,104
F_{12}^1	4	0,099
F_{11}^2	5	0,033
F_{21}^2	6	0,023
$F_{1,avg}^6$	7	0,016
F^7	8	0,010
$F_{2,avg}^6$	9	0,010
F_{22}^2	10	0,010
F_1^3	11	0,009
F_{12}^2	12	0,010
$F_{2,2}^4$	13	0,006
$F_{1,3}^4$	14	0,006
$F_{2,3}^4$	15	0,007
F_2^3	16	0,005
F_1^5	17	0,003
F_2^5	18	0,003
$F_{1,2}^4$	19	0,003
F_{avg}^8	20	0,003

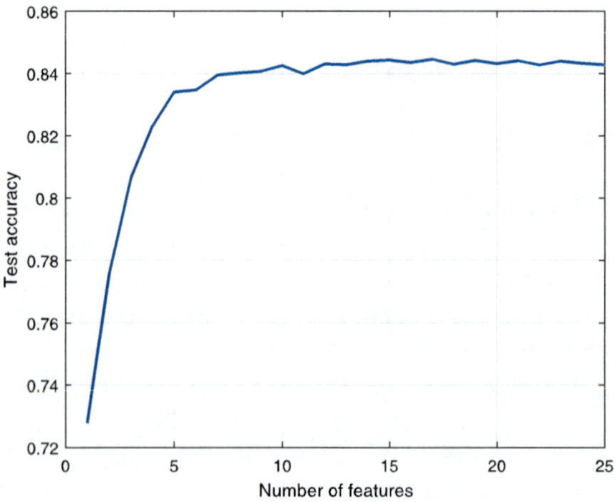

Fig. 1 Accuracy of the Gradient Boosting depending on the numbers of features retained by recursive elimination

Table 5 Cross validated scores for the Gradient Boosting model with selected features

Training/Test	Accuracy train (%)	Accuracy Test (%)	I_precision (%)	F_precision (%)	I_recall (%)	F_recall (%)	F1-score (%)
Same graphs	87.92	84.17	81.44	85.91	78.73	87.86	80.04
Different graphs	87.87	83.89	81.13	85.62	78.20	87.72	79.61

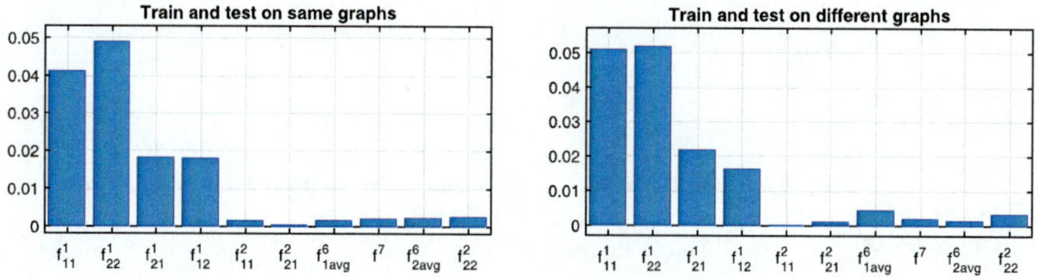

Fig. 2 Accuracy gap between a Gradient Boosting model trained with selected features and the same model trained excluding one of the features

excluding one of them. Again, we considered both the case with training and test set based on the same graphs and the case where different graphs are used. The features F^1_{hk}, for $h, k \in \mathcal{H}$, are the most relevant. The other features seem almost irrelevant, when taken out singularly. However, by comparing these results to those in Fig. 1 we conclude them to be in synergy: their contribution stacks up, finally providing overall an improvement of about 2%.

5 Conclusions

Our models, which are designed to predict the RCSPP feasibility by measuring features of the instance, prove effective. More in details, a supervised learning method, relying on the training of random forests by means of gradient boosting, gave the best results.

One of the main challenges was to design a set of features which is able to generalize from the specific structure of the training instances, and can be computed with a limited effort. In our experiments, no feature required more than milliseconds to compute on graphs with up to 5500 vertices. None of them depends on the presence of a specific graph structure.

We found the following outcome to be particularly relevant: by repeating our experiments with training and testing on different sets of graphs, accuracy kept above 84%.

Considering the perspective use of our models as fast approximate completion bound checks, we argue false Infeasible predictions to be the most critical ones. The

I_precision score deserves most attention: even if it results in values above 81% future research efforts in improving it might be pertinent.

The most promising way of integrating our models in exact methods is indeed their use as triggers for more computationally expensive checks.

For what concerns the embedding of our technique in column generation pricing, future investigations will extend our models to cover the *elementary* version of the problem, which allows for negative resource values, potentially leading to cycles.

References

1. Pugliese, L.D.P., Guerriero, F.: A Survey of Resource Constrained Shortest Path Problems: Exact Solution Approaches. Networks **62**, 183–200 (2013)
2. Toth, P., Vigo, D.: The Vehicle Routing Problem. In: MOS-SIAM Series on Optimization (2014)
3. Desaulniers, G., Desrosiers, J., Solomon, M.M.: Column Generation. Springer, New York (2005)
4. Martinelli, R., Pecin, D., Poggi, M.: Efficient elementary and restricted non-elementary route pricing. Eur. J. Oper. Res. **239**(1), 102–111 (2014)
5. Bezzi, D., Ceselli, A., Righini, G.: Automated tuning of a column generation algorithm. In: Kotsireas, I., Pardalos, P. (eds.) Learning and Intelligent Optimization. LION 2020. Lecture Notes in Computer Science, vol. 12096. Springer, Cham (2020). https://doi.org/10.1007/978-3-030-53552-0_21
6. Morabit, M., Desaulniers, G., Lodi, A.: Machine-learning-based arc selection for constrained shortest path problems in column generation. INFORMS J. Optim. **5**(2), 191–210 (2022). https://doi.org/10.1287/ijoo.2022.0082
7. Larose, D.T., Larose, C.D.: Data Mining and Predictive Analytics. In: Wiley Series on Methods and Applications in Data Mining. Wiley (2015)
8. 9th DIMACS Implementation Challenge—Shortest Paths. Challenge website: http://www.diag.uniroma1.it/~challenge9/
9. Hart, W.E., Watson, J., Woodruff, D.L.: Pyomo: modeling and solving mathematical programs in Python. Math. Program. Comput. **3**(3), 219–260 (2011)
10. Hart, W.E., et al.: Pyomo-Optimization Modeling in Python, vol. 67, 2nd edn. Springer Science & Business Media (2017)
11. Pedregosa, F., et al.: Scikit-learn: machine learning in Python. J. Mach. Learn. Res. **12**, 2825–2830 (2011)

Monte-Carlo Integration on a Union of Polytopes

Jonas Stübbe and Anne Remke

Abstract A new integration approach is presented that is tailored towards integrating over a union of polytopes with low coverage in high dimensions. It combines *Markov Chain Monte Carlo* and *Multiphase Monte Carlo* and takes advantage of the specific geometrical structure by directly sampling from it, which ensures scalability in higher dimensions. A feasibility study shows the efficiency of our method in comparison to the state-of-the-art approach *GSL VEGAS*. To showcase the specific strenght of the proposed method, integration is performed on a selected set of such multi-dimensional polytopes with low coverage.

1 Introduction

Verifying stochastic hybrid systems requires integrating the probability density function of random variables defined on a union of polytopes [14, 19, 20]. However, integrating over complex geometrical structures in large dimensions is usually not feasible analytically. Hence, mostly numerical solutions are applied, which sample uniformly from a rectangular overapproximation, e.g., *VEGAS* with stratified and importance sampling [16, 22], as implemented in the *GNU Scientific Library* [12]. Depending on the model, these method however yield excessive computation times in higher dimensions. Note, that *VEGAS* does not take advantage of the convexity of the polytopes computed in the reachability analysis.

Sampling from convex polytopes is well-investigated and methods like Markov-Chain Monte Carlo [2, 15, 18, 24, 25, 27, 28] and Multiphase Monte Carlo [6, 9–11] can be applied, efficiently. Other efficient sampling techniques build on a Markov chain, e.g., Ball Walk [13] or Billiard Walk [21]. A stable library for volume approximation and sampling of convex polytopes can be found in *VolEsti* [5]. Convergence of

J. Stübbe · A. Remke (✉)
University of Münster, Einsteinstr. 62, 48149 Münster, Germany
e-mail: anne.remke@uni-muenster.de

J. Stübbe
e-mail: jonas.stuebbe@uni-muenster.de

A. Brieden et al. (eds.), *Graphs and Combinatorial Optimization: from Theory to Applications*, AIRO Springer Series 13,
https://doi.org/10.1007/978-3-031-46826-1_12

Monte-Carlo *Hit-and-Run* methods is polynomial for convex polytopes [8], however can be arbitrarily slow for non-convex regions [1].

In 1984, Smith [25] proposed a method to sample from high dimensional non-trivial bodies called *rejection sampling* which samples from a bounding box instead which completetly contains the original body. However, the required number of samples grows exponentially for increasing dimensions, as the ratio between the volume of the original body and its bounding box decreases exponentially with increasing dimensions. When a union of polytopes is used, the corresponding bounding box is created based on the minimum and maximum point in each dimension over all polytopes, resulting in even lower coverage.

This paper proposes a Monte-Carlo integration *Multiphase Union Markov Chain Monte Carlo* (*MpU(MC)*^{2}) specifically tailored to the union of convex polytopes, which is in general not convex. *Multiphase Union Markov Chain Monte Carlo* addresses challenges which arise from a non-convex area of integration: First, we adapt coordinate *Hit-and-Run* to allow direct transitions between polytopes with an empty intersection. Rejection sampling is only used to obtain the initial sample, hence the *curse of dimensionality*, described above, does not impact performance as badly. All other samples are generated via our version of coordinate *Hit-and-Run* and are hence contained in the union of polytopes by construction. Second, the volume of the union of convex polytopes is approximated with a variant of *Multiphase Monte Carlo* that can be applied to non-convex regions. Then Monte-Carlo integration can be used to estimate a multi-dimensional integral over that union of convex polytopes.

Organization. Section 2 defines relevant preliminaries. Section 3 presents our approach of Monte-Carlo integration on the union of polytopes. Section 4 compares results obtained from *MpU(MC)*^{2}with *VEGAS*. Section 5 concludes the paper.

2 Preliminaries

Vectors are denoted as bold lowercase letters and matrices are denoted as bold uppercase letters. Sets of sets are denoted as calligraphic capital letters. Further, we define *intervals* as sets $T = \{v \in \mathbb{R} \mid l \leq v \leq u\}$ for some $l \in \mathbb{R}$ and $u \in \mathbb{R}$. Let $\mathcal{T}$ denote a set of intervals. To allow the definition of H-polytopes via the intersection of half-spaces, we first define half-spaces and hyperplanes. For two vectors $\mathbf{x}$ and $\mathbf{y}$, the scalar product is denoted by $\langle \mathbf{x}, \mathbf{y} \rangle \in \mathbb{R}$.

Definition 1 (*Half-space*) Given *normal vector* $\mathbf{n} \in \mathbb{R}^d$ and an offset $o \in \mathbb{R}$, a d-dimensional *half-space* h is defined as the set

$$h = \{\mathbf{x} \in \mathbb{R}^d | \langle \mathbf{n}, \mathbf{x} \rangle \leq o\}.$$

The set $\overline{h} = \{\mathbf{x} \in \mathbb{R}^d \mid \langle \mathbf{n}, \mathbf{x} \rangle = o\}$ is the *bounding hyperplane* of half-space h.

In the following, we consider polytopes in the so-called H-representation.

Definition 2 (*H-Polytope*) A H-polytope $P \subseteq \mathbb{R}^d$ is defined by the (bounded) intersection of m half-spaces

$$P := \{\mathbf{x} \in \mathbb{R}^d \mid \mathbf{A}\mathbf{x} \leq \mathbf{b}\},$$

with $\mathbf{A} \in \mathbb{R}^{m \times d}$, $\mathbf{b} \in \mathbb{R}^m$. The k-th row of $\mathbf{A}$ is refered to as $\mathbf{a}_k$ and the k-th half-space of the polytope as $\mathbf{a}_k\mathbf{x} \leq b_k$. Half-spaces are also called constraints.

The union of polytopes P_i, P_j is defined by $P_i \cup P_j = \{\mathbf{x} \in \mathbb{R}^d \mid \mathbf{x} \in P_i \vee \mathbf{x} \in P_j\}$, and the intersection is defined as $P_i \cap P_j = \{\mathbf{x} \in \mathbb{R}^d \mid \mathbf{x} \in P_i \wedge \mathbf{x} \in P_j\}$. Furthermore, $\mathcal{P} = \{P_i \mid 0 \leq i < n\}$ represents a set of polytopes. Let the size of $\mathcal{P}$ be denoted by $|\mathcal{P}| = n$. Here, $\bigcup_{P_i \in \mathcal{P}} P_i$ represents the union of polytopes P_i over the set $\mathcal{P}$.

H-Polytopes can be converted into other representations, e.g., vertex-oriented V-Polytopes. However, this transformation is NP-hard in general [3, 7]. In the following, we rely on the representation as H-polytopes.

Monte-Carlo integration is used to numerically approximate an integral. Sample points are drawn randomly from the integration region Ω and used to evaluate the integrand. We follow the definition in the *GNU scientific library* [12].

Definition 3 (*Monte-Carlo integration*) An multi-dimensional definite integral I is defined for $\mathbf{x} = (x_1, \ldots, x_d)$:

$$I = \int_\Omega f(\mathbf{x})d\mathbf{x},$$

with integrand $f : \mathbb{R}^d \rightarrow \mathbb{R}$ and $\Omega \subset \mathbb{R}^d$. The volume of Ω, denoted $V(\Omega) \in \mathbb{R}$ equals $V(\Omega) = \int_\Omega d\mathbf{x}$. With $N \in \mathbb{N}$ sample points $\mathbf{x_i} \in \mathbb{R}^d$, that are uniformly drawn from Ω, I can be approximated the integral as:

$$I \approx E(f; N) = \frac{V(\Omega)}{N} \sum_{i=1}^{N} f(\mathbf{x}_i). \tag{1}$$

Due to the strong law of large numbers, $E(f; N)$ converges to I for large N [23].

Let $\mathcal{N}(\mu, \sigma^2)$ denote the normal distribution with mean μ and variance σ^2. The function $g : \mathbb{R} \rightarrow \mathbb{R}$ then denotes the probability density function of $\mathcal{N}(\mu, \sigma^2)$ with:

$$g(x) = \frac{1}{\sigma\sqrt{2\pi}} \mathrm{e}^{-\frac{1}{2}(\frac{x-\mu}{\sigma})^2}.$$

This paper aims to integrate the joint density function of d i.i.d random variables, each distributed with $\mathcal{N}(\mu, \sigma^2)$ over Ω. Hence, we define $f(\mathbf{x}) = \prod_{l=1}^{d} g(x_l)$.

Rejection sampling [25] draws uniformly distributed samples from an overapproximating region $S \supseteq \Omega$ and $\Omega \subset \mathbb{R}^d$ instead of sampling over Ω directly. Each sample $\mathbf{x}$ that is contained in Ω will be distributed uniformly in Ω. If the sample generated in S is not contained in Ω, it is rejected. Usually, S is chosen as a region with a tight bound on Ω from which uniformly distributed samples can easily be drawn, e.g., a hyperrectangle. If $\Omega = \bigcup_{P_i \in \mathcal{P}} P_i$ is a union of polytopes, the hyperrectangle is given by the minimum and maximum value over all polytopes $P_i \in \mathcal{P}$ in every dimension, further denoted as $S_{\mathcal{P}}$.

In the following, we use *Markov Chain Monte Carlo* in the variant *Hit-and-Run* as introduced by Smith [26] in 1984. *Hit-and-Run* draws samples from a homogenous continuous-state Markov chain in discrete time, i.e., $\{X_n \mid n \geq 0\}$ with $n \in \mathbb{N}$. Such a Markov is chain is called a *Random Walk* and defined as a family of independently and identically distributed random variables, which take values from $\mathbb{R}^d$. For a set $\Omega \subset \mathbb{R}^d$, *Hit-and-Run* generates a direction vector $\mathbf{v} \in \mathbb{R}^d$, that is uniformly distributed on the direction set $D \subseteq \mathbb{R}^d$, depending on the used *Hit-and-Run* method.

Let $l = \{\mathbf{x} + \lambda \cdot \mathbf{v} \mid \lambda \in \mathbb{R}\}$ be the line defined by $\mathbf{x}$ and $\mathbf{v}$. A new point $\mathbf{x}'$ is then sampled uniformly distributed from the part of l that is included in Ω.

For convex spaces Ω the Markov chain converges to approximately the stationary distribution [25]. Here, the convergence speed is specified as mixing time [18], which indicates the required number of steps in the Random Walk until obtaining points which are approximately stationary.

Hit-and-Run is considered to be one of the most efficient *Markov Chain Monte Carlo* sampler currently available for generating asymptotically uniform points in convex sets due to its polynomial convergence time [17, 18]. As hit-and-run sampling algorithms produce points that are uniformly distributed on a convex space, they are very well suited for sampling polytopes.

Coordinate Directions Hit-and-Run (*CDHR*) [9] is a form of *Hit-and-Run*. Starting from a point $\mathbf{x} \in \mathbb{R}^d$ inside the polytope $P = \{\mathbf{x} \in \mathbb{R}^d \mid \mathbf{A}\mathbf{x} \leq \mathbf{b}\}$, $\mathbf{v}$ is uniformly sampled at random as one of the coordinate axes. Afterwards, $\mathbf{v}$ and $\mathbf{x}$ are used to calculate the distance $\lambda \in \mathbb{R}$ to the bounding hyperplane of every half-space of P. The scalars λ^+ and λ^- result from maximizations of λ over each half-space [9]:

Definition 4 (*Minimum distance to bounding hyperplane*) Given a convex polytope P_i with m_i half-spaces, a point $\mathbf{x} \in P_i$ and a direction vector $\mathbf{v}$. Let $\lambda^+, \lambda^- \in \mathbb{R}$ be the minimum distance between point $\mathbf{x}$ and a bounding hyperplane $\overline{h}$ along $\mathbf{v}$ in both directions. λ^+ and λ^- are calculated by maximizing the distance $\lambda \in \mathbb{R}$ to the bounding hyperplane of each half-space with

$$\lambda^+ = \max\{\lambda \mid \forall k, 0 \leq k < m_i : \mathbf{a}_k(\mathbf{x} + \lambda\mathbf{v}) \leq b_k\}, \text{ and}$$
$$\lambda^- = \max\{\lambda \mid \forall k, 0 \leq k < m_i : \mathbf{a}_k(\mathbf{x} - \lambda\mathbf{v}) \leq b_k\}.$$

The interval T_i is then defined by $T_i = [\lambda^-, \lambda^+]$.

We now uniformly sample a value $t \in T_i$ and as a result, a new point $\mathbf{x}' = \mathbf{x} + t \cdot \mathbf{v}$ is obtained, which by construction is contained in the polytope. The iterative procedure is continued until a predefined number of sample points has been generated. The generation of three sample points with *Coordinate Directions Hit-and-Run* on a polytope is illustrated in Fig. 1 from left to right: starting from a point $\mathbf{x}$, first a direction vector $\mathbf{v}$ is drawn and the line segment $l = \{\mathbf{x} + t \cdot \mathbf{v} \mid t \in T_i\}$ created. From l, a new sample point $\mathbf{x}'$ is drawn uniformly.

Note that the sample points correspond to the values taken by the random variables of the Markov chain $\{X_n | n \geq 0\}$, with $X_n \in \mathbb{R}^d$.

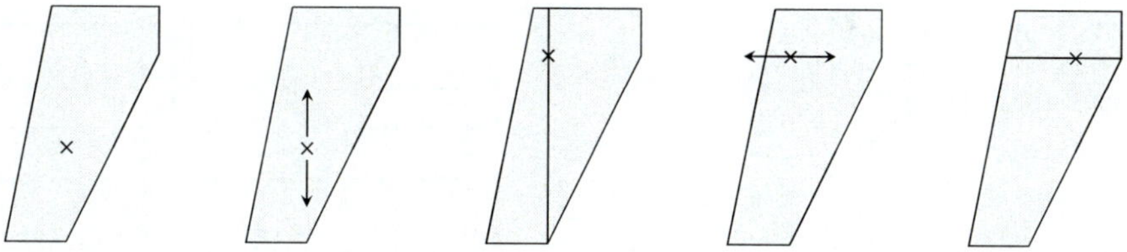

Fig. 1 *Coordinate Directions Hit-and-Run* on a polytope

3 Monte-Carlo Integration on a Union of Polytopes

Hit-and-Run Monte-Carlo methods are not restricted to convex polytopes. Smith even proves in [27] that *Hit-and-Run* algorithms converge to approximately the uniform distribution when applied to an open subset of $\mathbb{R}^d$. However, their mixing time, i.e., the bound on the number of samples required for convergence, can be arbitrarily large for non-convex regions [1].

Figure 2 shows an examplary run of traditional Markov chain *Hit-and-Run* on a (non-convex) union of three polytopes. Here, starting in P_0, the Markov chain can only directly transition into polytopes with a non-empty intersection with P_0.

In the following, we detail two major improvements of *MpU(MC)*^{2}, which allow the transition to unions of polytopes. First, we allow direct transitions from one polytope to another one that is part of that union, however has an empty intersection with the current one. This is illustrated in Fig. 3.

Second, the volume of the union is computed with a variant of *Multiphase Monte Carlo*, and used to approximate the integral in Eq. 1. As the intersection of any two polytopes in the union is not necessarily empty (e.g., see Fig. 2) the volume of that union does not simply equal the sum of the individual polytope's volumes.

Section 3.1 extends coordinate *Hit-and-Run* to a union of convex polytopes and allows direct transitions between two polytopes with empty intersection. Section 3.2 discusses how the volume of the union is approximated via *Multiphase Monte Carlo*.

Fig. 2 A possible path from P_0 to P_2 results in three intermediate points in P_1

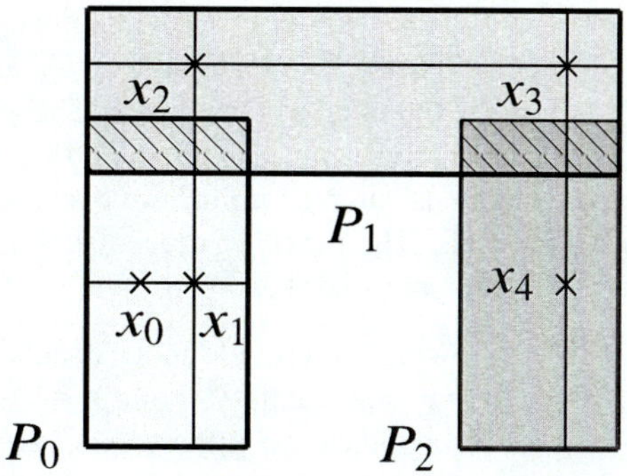

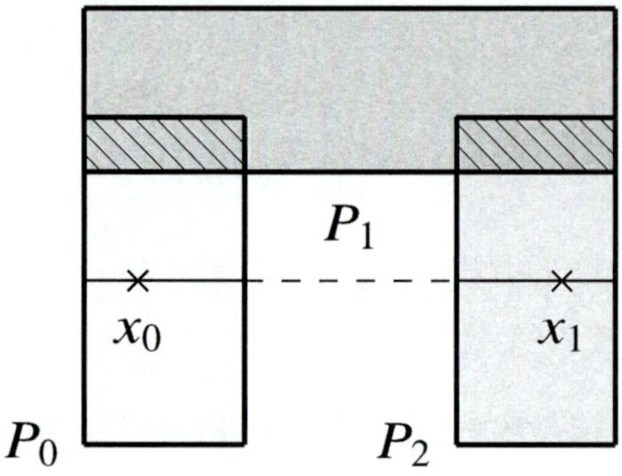

Fig. 3 A possible path from P_0 to P_2 is possible without intermediate points in P_1

3.1 Extending Markov Chain Monte Carlo

The union of convex polytopes $\bigcup_{P_i \in \mathcal{P}} P_i$ is in general not convex. Hence, the mixing-time of *CDHR* could be arbitrarily large [1]. Intuitively, this is due to the fact that *CDHR* only generates successor sample points within the same polytope P_i or in any polytope P_j with a non-empty intersection $P_i \cap P_j \neq \emptyset$.

Figure 2 illustrates a sequence of sampling points starting in $\mathbf{x}_0$, where the Markov chain requires sampling from polytope P_1 before samples in P_2 can be constructed. In the following, *Markov Chain Monte Carlo* is extended by allowing transitions from polytope P_0 to P_2 without intermediate samples in P_1, as shown in Fig. 3. We require the set $\mathcal{P}$ to include at least two polytopes, i.e., $|\mathcal{P}| \geq 2$. The iterative computation of sample points with *MpU(MC)*^{2}then works as follows:

(1) To initialize one execution of *MpU(MC)*^{2}, we perform rejection sampling from the bounding box of the union of polytopes, denoted $S_{\mathcal{P}}$, until a point $\mathbf{x} \in \mathbb{R}^d$ is sampled, which is contained in Ω.
(2) A direction vector $\mathbf{v}$ is uniformly sampled at random as one of the coordinate axes. Let $\mathcal{T}$ denote a set of intervals, which is initially empty.
(3) Let $\mathbf{x}$ be contained in polytope P_i, then the corresponding interval T_i is created according to Definition 4 and added to the set of intervals $\mathcal{T}$.
(4) Let $P_j \in \mathcal{P}$ with $P_i \neq P_j$ be another polytope. Then either $\mathbf{x} \in P_j$ or $\mathbf{x} \notin P_j$ applies. If $\mathbf{x} \in P_j$, create interval T_j according to Definition 4 and add to $\mathcal{T}$. If $\mathbf{x} \notin P_j$, the interval T_j is created according to Definition 5 and added to $\mathcal{T}$.
(5) After iterating over each polytope $P_i \in \mathcal{P}$, a new point is generated with $\mathbf{x} + t \cdot \mathbf{v}$, where t is sampled uniformly over the set $\mathcal{T}$ of all obtained one-dimensional intervals. The process is repeated from step 2., until a predefined number of sample points has been generated.

In the following we provide more details on step 4. Recall that a polytope P_j is defined by m_j constraints. A point $\mathbf{x} \in P_j$ then satisfies every constraint of P_j. In contrast for a point $\mathbf{x} \notin P_j$, there exists at least one constraint in P_j with $\mathbf{a}_q \mathbf{x} > b_q$ for $0 \leq q < m_j$. For such a sample point, the goal is to find the interval T_j such that for all $t \in T_j$, $\mathbf{a}_k(\mathbf{x} + t \cdot \mathbf{v}) \leq b_k$ is satisfied for all $0 \leq k < m_j$. As it is previously unknown how many constraints are violated by a point $\mathbf{x} \notin P_j$, the interval T_j requires a more elaborate computation than the one explained in Definition 4. Let $sgn : \mathbb{R} \to \{+, -, 0\}$ denote the mapping of each value of the real numbers to its respective sign. Recall that a half-space h is well defined by $\mathbf{n}$ and o. We now introduce two functions which are applied to every half-space h of P_j:

$$dir\colon \mathbb{R}^d \times \mathbb{R}^d \to \{+, -, 0\} \text{ with } dir(\mathbf{n}, \mathbf{v}) = sgn(\langle \mathbf{n}, \mathbf{v} \rangle), \text{ and}$$

$$sat\colon \mathbb{R} \times \mathbb{R}^d \times \mathbb{R}^d \to \{+, -, 0\} \text{ with } sat(o, \mathbf{n}, \mathbf{x}) = sgn(o - \langle \mathbf{n}, \mathbf{x} \rangle).$$

Here, *dir* specifies, whether the constraint's normal vector $\mathbf{n}$ and the direction vector $\mathbf{v}$ point in the same direction. It returns $+$, if both vectors point in the same direction and $-$ if they point in different directions. In case $\mathbf{v}$ and $\mathbf{n}$ are parallel, *dir* returns 0. Using the constraint's normal vector $\mathbf{n}$ and offset o, *sat* returns $+$, if a point $\mathbf{x}$ satisfies the constraint i.e., $\mathbf{x} \in h$. It returns $-$, if $\mathbf{x} \notin h$ and zero if $\mathbf{x} \in \overline{h}$.

Further, we use *dir* and *sat*, to split the line $l = \left\{\mathbf{x}' \mid \mathbf{x}' = \mathbf{x} + \lambda \cdot \mathbf{v}, \lambda \in \mathbb{R}_{>0}\right\}$ into two rays $l^+ = \left\{\mathbf{x}' \mid \mathbf{x}' = \mathbf{x} + \lambda \cdot \mathbf{v}, \lambda \in \mathbb{R}^+_{>0}\right\}$ and $l^- = \left\{\mathbf{x}' \mid \mathbf{x}' = \mathbf{x} - \lambda \cdot \mathbf{v}, \lambda \in \mathbb{R}^+_{>0}\right\}$.

As each individual polytope P_j is convex and $\mathbf{x} \notin P_j$, if the set $l \cap P_j$ is non-empty, the set of points $l \cap P_j$ is either completely included in ray l^+ or in ray l^-. If *dir* or *sat* is zero, none of the rays l^+, l^- intersects the bounding hyperplane. Otherwise, if *dir* equals *sat*, the ray l^+ intersects the bounding hyperplane, and if *dir* does not equal *sat*, l^- intersects the bounding hyperplane.

Figure 4 illustrates the four different combinations of *dir* and *sat*, where $\mathbf{x}$ is indicated by x, the arrow originating in $\mathbf{x}$ specifies the sampled direction vector $\mathbf{v}$, and for the currently considered halfspace h (indicated in red) its normal vector $\mathbf{n}$ is shown. In cases **(a)**, and **(b)** the bounding hyperplane $\overline{h}$ is intersected by l^+ and in cases **(c)** and **(d)** it is intersected by l^-.

Let $\mathcal{H}_i = \{\mathbf{a}_k\mathbf{x} \leq b_k \mid 0 \leq k < m_i\}$ be the set of all half-spaces for polytope P_i then $L^+ = \{h \mid h \in \mathcal{H}_i \wedge l \cap h \subseteq l^+\}$ is the set of constraints h in polytope P_i for which the intersection of l and h is completely included in ray l^+. Further let $L^- = \{h \mid h \in \mathcal{H}_i \wedge l \cap h \subseteq l^-\}$ be the set of constraints h in polytope P_i for which the intersection of l and h is completely included in ray l^-.

As polytope P_i is convex and $\mathbf{x} \notin P_i$, the interval T_i is created either by the set of constraints L^+ or L^-. Therefore, we consider both sets separately: We first consider the set L^+ (compare **(a)**, **(b)** in Fig. 4). If the set $l^+ \cap P_i$ is non-empty, we define the interval T_i using only constraints in L^+, as follows:

Definition 5 (*Line segment boundaries for* $\mathbf{x} \notin P_i$) Given a polytope P_i with m_i half-spaces denoted by the set $\mathcal{H}_i = \{\mathbf{a}_k\mathbf{x} \leq b_k \mid 0 \leq k < m_i\}$, a point $\mathbf{x} \in \mathbb{R}^d$, $\mathbf{x} \notin P_i$ and a direction vector $\mathbf{v}$. Let $L^+ \subseteq H_i$ be a set of constraints with $|L^+| = m_i' \leq m_i$. Then $L^+_{in} \subseteq L^+$ and $L^+_{out} \subseteq L^+$ is defined by

$$L^+_{in} = \{h \mid h \in L^+ \wedge \mathbf{x} \notin h\}, \text{ and } L^+_{out} = \{h \mid h \in L^+ \wedge \mathbf{x} \in h\}.$$

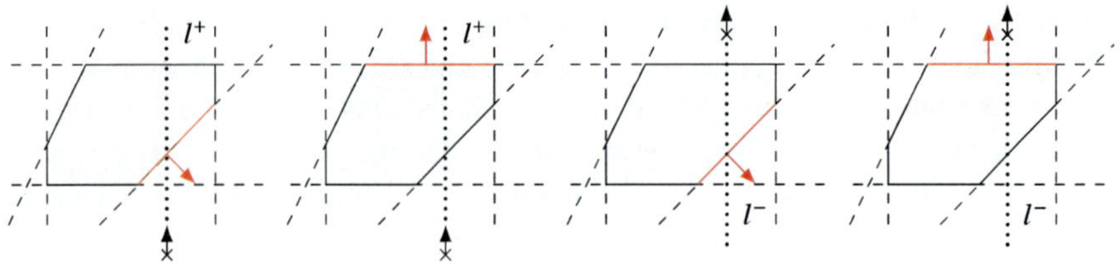

Fig. 4 Classification of constraints according to the values of $dir(\mathbf{n}, \mathbf{v})$ and $sat(o, \mathbf{n}, \mathbf{x})$

Further, let $\lambda^{in}, \lambda^{out} \in \mathbb{R}$ be two distances defined by

$$\lambda^+_{in} = \max\{\lambda \mid h \in L^+_{in} \wedge \langle \mathbf{n}, \mathbf{x} + \lambda\mathbf{v} \rangle = o\}, \text{ and}$$
$$\lambda^+_{out} = \min\{\lambda \mid h \in L^+_{out} \wedge \langle \mathbf{n}, \mathbf{x} + \lambda\mathbf{v} \rangle = o\}.$$

Then λ^+_{in} denotes the maximal distance along $\mathbf{v}$ between point $\mathbf{x}$ and a bounding hyperplane $\bar{h}$, for $h \in L^+_{in}$. Further, λ^+_{out} denotes the minimal distance along $\mathbf{v}$ between point $\mathbf{x}$ and a bounding hyperplane $\bar{h}$, for $h \in L^+_{out}$. The interval T_i is then defined by $T_i = [\lambda^+_{in}, \lambda^+_{out}]$.

The above definition is adapted to L^-, using

$$L^-_{in} = \{h \mid h \in L^- \wedge \mathbf{x} \notin h\}, \text{ and } L^-_{out} = \{h \mid h \in L^- \wedge \mathbf{x} \in h\},$$

and $T_i = [-\lambda^-_{out}, \lambda^-_{in}]$ is defined by:

$$\lambda^-_{in} = \max\{\lambda \mid h \in L^-_{in} \wedge \langle \mathbf{n}, \mathbf{x} - \lambda\mathbf{v} \rangle = o\}, \text{ and}$$
$$\lambda^-_{out} = \min\{\lambda \mid h \in L^-_{out} \wedge \langle \mathbf{n}, \mathbf{x} - \lambda\mathbf{v} \rangle = o\}.$$

After the extensions made to Markov Chain Monte Carlo, the procedure still satisfies the definition of Mixing Algorithms by Smith [25], therefore results in a sequence of sample points that is uniformly distributed in Ω.

3.2 Extending Multiphase Monte Carlo

Although a variety of Monte-Carlo algorithms for volume approximation exist, their majority can only be applied to convex bodies. More general approaches, e.g., based on partitioning into simplices, suffer from scalability issues. In the following we show how *Multiphase Monte Carlo* approximates the volume $V(\Omega)$ of the union over a finite set of convex polytopes $\Omega = \bigcup_{P_i \in \mathcal{P}} P_i$.

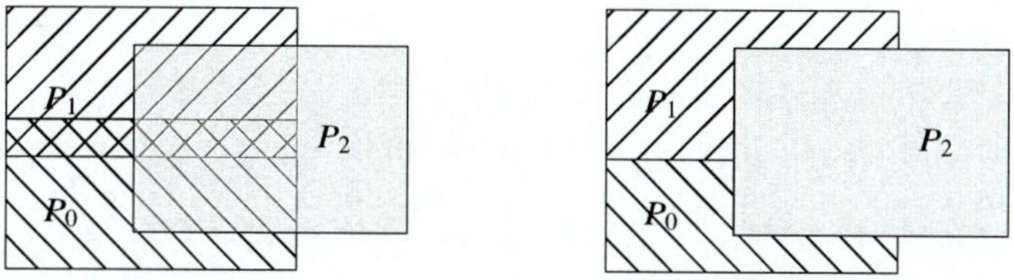

Fig. 5 Union of polytopes P_1, P_2, P_3 (left) and the unique assignment of the non-empty intersections to the polytope with the largest index (right)

First, we apply *Multiphase Monte Carlo* to each P_i individually, and approximate its volume $V(P_i)$ as V_i. Then we ensure that the volumes of the individual polytopes are combined such that non-empty intersections are taken into account, correctly. We attribute the non-empty intersection of polytopes P_i and P_j to the polytope with the higher index. Hence, every polytope P_j maintains the volume of its non-empty intersection with polytope P_i for $i < j$. Figure 5 illustrates the concept for the union of three polytopes (left) and the resulting attribution of intersections to the polytope with the highest index (right).

Instead of considering the union of convex polytopes P_i with potentially non-empty intersections, we take the union of non-convex polytopes $\hat{P}_i$ with empty intersections, i.e., $\hat{P}_i \cap \hat{P}_j = \emptyset$ for all $i \neq j$ and $i, j \leq |\mathcal{P}|$, formalized as:

$$\Omega = \bigcup_{P_i \in \mathcal{P}} P_i = \bigcup_{P_i \in \mathcal{P}} \hat{P}_i, \text{ with } \hat{P}_i = \{\mathbf{x} \in P_i \mid \forall P_j \in \mathcal{P}, i < j.\mathbf{x} \notin P_j\}.$$

Second, we sample N points from Ω with *MpU(MC)*^{2}, resulting in the set of samples $\Phi_N = \{\mathbf{x}_i \mid 0 \leq i \leq N\}$ which are uniformly distributed over Ω. We then estimate the volume of each $\hat{P}_i$ using the samples in Φ_N and the indicator function $\mathbb{1}_{P_i}(\mathbf{x})$ which returns one if $\mathbf{x} \in P_i$ and zero, otherwise. The number of samples from Φ_N contained in the polytope P_i is then given as $\sum_{\mathbf{x} \in \Phi_N} \mathbb{1}_{P_i}(\mathbf{x})$. For simplicity we write $\sum_{\Phi_n} \mathbb{1}_{P_i}$. For every polytope $P_i \in \mathcal{P}$, we define ρ_i, which is used to reduce the estimated volume V_i to $V_{\hat{P}_i}$:

$$\rho_i = \frac{\sum_{\Phi_n} \mathbb{1}_{\hat{P}_i}}{\sum_{\Phi_n} \mathbb{1}_{P_i}}, \text{ and } V_{\hat{P}_i} = \rho_i \cdot V_i.$$

Hence, the volume of Ω is approximated as $V_\Omega = \sum_{i=1}^{|\mathcal{P}|} \rho_i \cdot V_i$. This allows to estimate the integral from Eq. 1, using $V(\Omega) = V_\Omega$.

4 Evaluation

We compare results obtained by *MpU(MC)*^{2}with the state-of-the-art approach *GSL VEGAS* for pseudo-randomly created sets of polytopes $\mathcal{P}$ in 15 and 20 dimensions. All results were obtained on a *Nehalem-C @2.25 GHz* processor. We compute the probability resulting from Eq. 1 for the joint density function $f = \prod_{l=1}^{d} g(x_l)$, where $g(x)$ equals the normal distribution $\mathcal{N}(\mu, \sigma^2)$ with parameters $\mu = 10$, $\sigma = 5$ in every dimension.

Results obtained by *MpU(MC)*^{2}and *VEGAS* are illustrated as boxplots indicating median and variation for a union of polytopes in 15-dimensions in Fig. 6 and for 20 dimensions in Fig. 7. The whiskers are chosen as 1.5 times the range between the upper and the lower quartile. Table 1 summarizes the corresponding number of runs, the average computation time of runs and the median for all performed integrations.

For $d = 15$, *MpU(MC)*^{2}and *VEGAS* compute a comparable median in a similar time, while *VEGAS* achieves a smaller variation, as indicated in the boxplots. However, for $d = 20$, *MpU(MC)*^{2}clearly outperforms *VEGAS*, as the resulting boxplots indicate a very small variation, while *VEGAS* results in a large variation. Note that the median computed by *MpU(MC)*^{2}and *VEGAS* differ considerably. As no ground truth exists for these computations, we consider the scatter plots illustrating the computed probability over the computation time for $d = 15$ in Fig. 8 and for $d = 20$ in Fig. 9 for every run performed.

Recall that *MpU(MC)*^{2}is an iterative approach which computes a fixed number of samples. In contrast *VEGAS* uses a convergence criterion $\chi^2 \leq 0.5$ and required between one and seven iterations per run. Figures 8 and 9 show that runs with

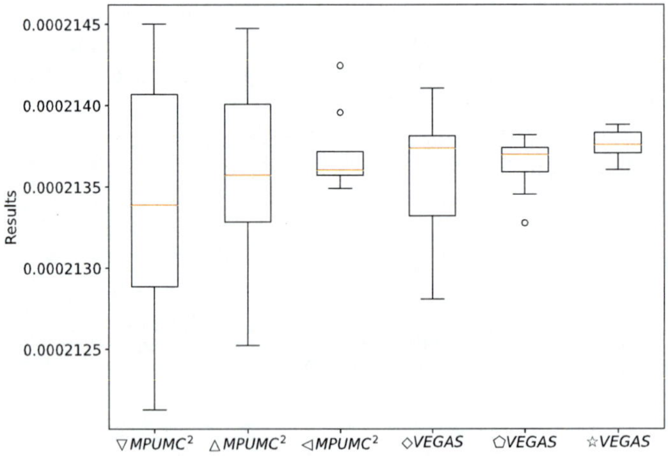

Fig. 6 *MpU(MC)*^{2}with $N_{\triangledown} = 1e{+}8$, $N_{\triangle} = 2, 5e8$ and $N_{\triangleleft} = 5e{+}8$, *VEGAS* with $N_{\diamond} = 4, 3e{+}8$, $N_{\pentagon} = 9, 3e{+}8$ and $N_{\star} = 2, 4e{+}9$, $d = 15$

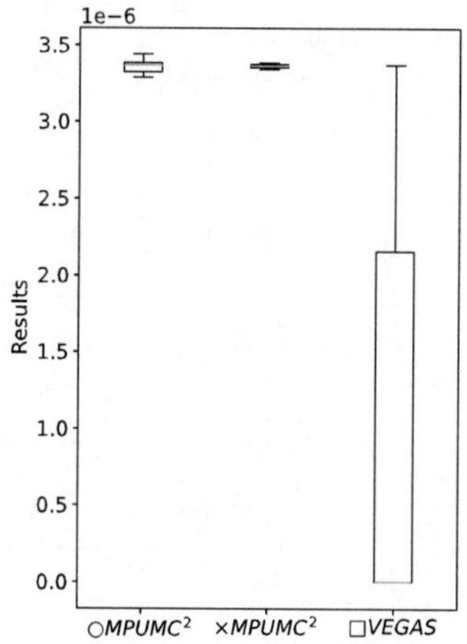

Fig. 7 *MpU(MC)²*with $N_{\bigcirc} = 1e{+}7$ and $N_{\times} = 1e{+}8$, *VEGAS* with $N_{\square} = 3,5e{+}9$, $d = 20$

Table 1 Computation times, number of runs and median obtained in dimensions 15 and 20 with *MpU(MC)*² and *VEGAS*, respectively

	*MpU(MC)*² ▽	*MpU(MC)*² △	*MpU(MC)*² ◁	*MpU(MC)*² ○	*MpU(MC)*² ×	*VEGAS* ◇	*VEGAS* ⬠	*VEGAS* ☆	*VEGAS* □
Dimension	15	15	15	20	20	15	15	15	20
Time(h)	1.79	4.45	9.01	4.73	12.98	2.14	4.35	9.46	27.76
# runs	7	7	8	10	10	7	8	7	10
Median (e+4)	2.134	2.136	2.136	0.034	0.034	2.137	2.137	2.138	9.485e-10

*MpU(MC)*²always have the same computation time for a fixed number of samples, while the computation time of *VEGAS* depends on the number of iterations. For $d = 15$ *VEGAS* computes a similar probability for different number of iterations.

For $d = 20$, seven runs of *VEGAS* terminate after one or two iterations, i.e., 10h or 20h, and the resulting probability is too small. The three runs of *VEGAS* with larger computation times take 4, 6 and 7 iterations and result in a very similar probability as computed by *MpU(MC)*². These results indicate, that *VEGAS* was started with not enough samples. However, the sample size of *VEGAS* in 20 dimensions was already set to $3,5e{+}9$, and the corresponding computation time is excessive.

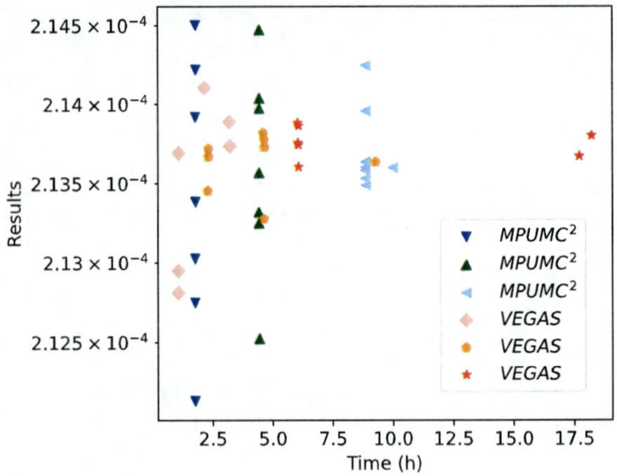

Fig. 8 Relation of time and result, $d = 15$

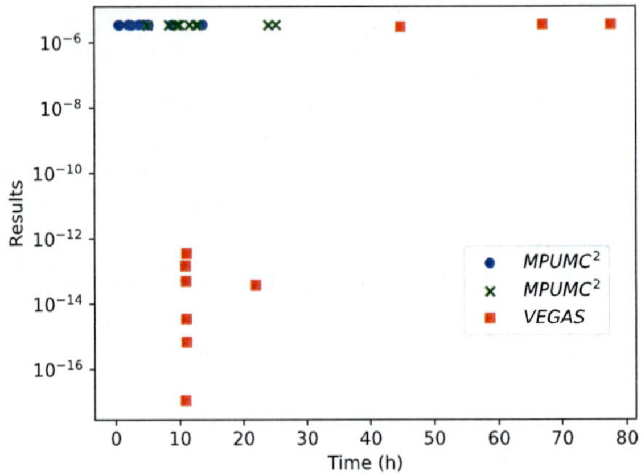

Fig. 9 Relation of time and result, $d = 20$

5 Conclusion

The proposed integration method *MpU(MC)*^{2}is able to provide very good approximations of multi-dimensional integrals over the union of convex polytopes, especially in high dimensions. The results of the evaluation show that *VEGAS* as implemented in the *GNU* library, is an efficient integration strategy for up to 15 dimensions. In 20 dimensions *MpU(MC)*^{2}clearly outperforms *VEGAS* on the randomly created union of polytopes considered. Note that due to excessive computation times, we do not have the same number of runs for all computations in 15 dimensions. Instead, the available computational power was focused on computations in 20 dimensions.

This work is funded through the DFG grant 471367371. To ensure repeatability, the source code and the generated sets of polytopes can be found at https://zivgitlab.uni-muenster.de/ag-sks/tools/realyst.

Future work aims at proving that the mixing time of *MpU(MC)*^{2} is polynomial and to include *MpU(MC)*^{2} into tools for verification of stochastic hybrid systems.

References

1. Abbasi-Yadkori, Y., Bartlett, P., Gabillon, V., Malek, A.: Hit-and-run for sampling and planning in non-convex spaces. In: 20th International Conference on Artificial Intelligence and Statistics, Proceedings of Machine Learning Research, vol. 54, pp. 888–895 (2017)
2. Asmussen, S., Glynn, P.: A new proof of convergence of MCMC via the ergodic theorem. Stat. Probab. Lett. **81**, 1482–1485 (2011)
3. Barber, C.B., Dobkin, D.P., Huhdanpaa, H.: The quickhull algorithm for convex hulls. ACM Trans. Math. Softw. **22**(4), 469–483 (1996)
4. Bellman, R.E.: Dynamic Programming. Princeton University Press, Princeton (2021)
5. Chalkis, A., Fisikopoulos, V.: Volesti: volume approximation and sampling for convex polytopes in R. R J. **13**(2), 561 (2021)
6. Cousins, B., Vempala, S.: A practical volume algorithm. Math. Program. Comput. **8**(2), 133–160 (2016)
7. Dyer, M.E.: The complexity of vertex enumeration methods. Math. Oper. Res. **8**(3), 381–402 (1983)
8. Dyer, M.E., Frieze, A.M.: Random walks, totally unimodular matrices, and a randomised dual simplex algorithm. Math. Program. **64**, 1–16 (1994)
9. Emiris, I.Z., Fisikopoulos, V.: Efficient random-walk methods for approximating polytope volume. In: 13th Annual Symposium on Computational Geometry, pp. 318–327. ACM (2014)
10. Ge, C., Ma, F.: A fast and practical method to estimate volumes of convex polytopes. In: Frontiers in Algorithmics, pp. 52–65. Springer, Berlin (2015)
11. Ge, C., Ma, F., Zhang, P., Zhang, J.: Computing and estimating the volume of the solution space of SMT(LA) constraints. Theoret. Comput. Sci. **743**, 110–129 (2018)
12. Gough, B.: GNU scientific library reference manual. Network Theory Ltd. (2009)
13. Hastings, W.K.: Monte Carlo sampling methods using Markov chains and their applications. Biometrika **57**(1), 97–109 (1970)
14. Hüls, J., Pilch, C., Schinke, P., Niehaus, H., Delicaris, J., Remke, A.: State-space construction of hybrid petri nets with multiple stochastic firings. ACM Trans. Model. Comput. Simul. **31**(3), 13:1–13:37 (2021)
15. Kiatsupaibul, S., Smith, R.L., Zabinsky, Z.B.: An analysis of a variation of hit-and-run for uniform sampling from general regions. ACM Trans. Model. Comput. Simul. **21**(3) (2011)
16. Lepage, G.P.: Adaptive multidimensional integration: VEGAS enhanced. J. Comput. Phys. **439**, 110386 (2021)
17. Lovasz, L., Vempala, S.: Fast algorithms for logconcave functions: sampling, rounding, integration and optimization. In: 47th Annual IEEE Symposium on Foundations of Computer Science, pp. 57–68 (2006)
18. László, L.: Hit-and-run mixes fast. Math. Program., Ser. B **86**, 443–461 (1999)
19. Pilch, C., Hartmanns, A., Remke, A.: Classic and non-prophetic model checking for hybrid petri nets with stochastic firings. In: 23rd International Conference on Hybrid Systems: Computation and Control. ACM (2020)
20. Pilch, C., Schupp, S., Remke, A.: Optimizing reachability probabilities for a restricted class of stochastic hybrid automata via Flowpipe-construction. In: Quantitative Evaluation of Systems, pp. 435–456. Springer, Berlin (2021)
21. Polyak, B., Gryazina, E.: Billiard walk—a new sampling algorithm for control and optimization. IFAC Proc. Vol. **19**, 6123–6128 (2014)
22. Press, W.H., Farrar, G.R.: Recursive stratified sampling for multidimensional Monte Carlo integration. Comput. Phys. **4**(2), 190 (1990)

23. Ross, S.M.: Introduction to probability models. Academic Press (2007)
24. Simonovits, M.: How to compute the volume in high dimension? Math. Program. **97**(1), 337–374 (2003)
25. Smith, R.L.: Efficient Monte Carlo procedures for generating points uniformly distributed over bounded regions. Oper. Res. (1984)
26. Smith, R.L.: Efficient Monte Carlo procedures for generating points uniformly distributed over bounded regions. Oper. Res. **32**(6), 1296–1308 (1984)
27. Smith, R.L.: The hit-and-run sampler: a globally reaching Markov chain sampler for generating arbitrary multivariate distributions. In: Conference on Winter Simulation, pp. 260–264. IEEE CS (1996)
28. Zabinsky, Z.B., Smith, R.L.: Hit-and-Run Methods, pp. 721–729. Springer, Berlin (2013)

Achieving Long-Term Fairness in Submodular Maximization Through Randomization

Shaojie Tang, Jing Yuan, and Twumasi Mensah-Boateng

Abstract Submodular function optimization is applied in ML and data analysis, including diverse dataset summarization. Fairness-aware algorithms are essential for handling sensitive attributes. Our research investigates the problem of maximizing a monotone submodular function while adhering to constraints on the expected number of selected items per group. Our goal is to compute a distribution over feasible sets, and to achieve this, we develop a series of approximation algorithms.

1 Introduction

A set function is referred to as submodular if it follows the principle of diminishing returns, where adding an item to a larger set yields a smaller benefit. This concept is applied in various real-world scenarios such as feature selection [9], where the goal is to select the most relevant features from a large pool of potential features to use in a machine learning model; active learning [14, 20, 21], where the goal is to choose a set of instances for a machine learning model to learn from; exemplar-based clustering [10], where the goal is to choose a set of exemplars to represent a set of data points; influence maximization in social networks [22, 24], where the goal is to choose a set of individuals to target in order to maximize the spread of information or influence in a network; as well as recommender system [12] and diverse data summarization [18]. The goal of submodular optimization is to choose a set of items that optimizes a submodular function while satisfying constraints such as size limitations, matroid requirements, or knapsack restrictions.

S. Tang (✉)
Naveen Jindal School of Management, The University of Texas at Dallas, Richardson, USA
e-mail: shaojie.tang@utdallas.edu

J. Yuan · T. Mensah-Boateng
Department of Computer Science and Engineering, University of North Texas, Denton, USA
e-mail: jing.yuan@unt.edu

T. Mensah-Boateng
e-mail: twumasimensah-boateng@my.unt.edu

A. Brieden et al. (eds.), *Graphs and Combinatorial Optimization: from Theory to Applications*, AIRO Springer Series 13,
https://doi.org/10.1007/978-3-031-46826-1_13

In practice, items or individuals are often grouped based on attributes such as gender, race, age, religion, or other factors. However, if not properly monitored, existing algorithms may display bias and result in an over- or under-representation of certain groups in the final selected set. To address this issue, we propose the study of long-term fair submodular maximization problem. The aim is to randomly choose a subset of items that optimizes a submodular function, such that the expected number of selected items from each group falls within the desired range. This approach ensures that the final selection of items is not only optimized, but also equitable, providing a fair representation of all groups in the long term.

Formally, we consider a set V of items, which are divided into m (not necessarily disjoint) groups: $V_1, V_2, \ldots, V_m$ with items in each group sharing similar attributes (e.g., race). To ensure fairness, a randomized item selection algorithm must satisfy the following criteria for all groups $t \in [m]$ where $[m] = \{1, 2, \ldots, m\}$: the *expected* number of selected items from group V_t must be within the range of $[\alpha_t, \beta_t]$, where α_t and β_t are arbitrary parameters that may differ across groups; moreover, the number of chosen items must *always* stay below a cardinality constraint of b. To put it simply, a fair randomized solution must meet two important requirements [3]: (a) restricted dominance, which means the proportion of items from each group must be within a certain limit, and (b) minority protection, which means the proportion of items from each group must not fall below a certain limit. Our fairness notation has gained significant recognition in the academic world and it has been adopted in various studies, including multi-winner voting systems [7], fair recommendation systems [13], and matroid-constrained optimization problems [8]. In fact, this notation is capable of capturing other fairness definitions such as statistical parity [11], the 80%-rule [4], and proportional representation [16].

In contrast to the majority of prior studies on fairness-aware algorithm design [7, 8, 13, 23, 25], which focus on finding a *fixed* solution set, our objective is to compute a randomized solution that can, on average, satisfy the group fairness constraints. This approach provides more flexibility in meeting the fairness requirements. Take fairness-aware product recommendations as an example. The objective is to suggest a set of products to online consumers while ensuring that each group of sellers, such as male and female sellers, is expected to have at least one of their products recommended. Due to limited display space, suppose we can only display one product to the consumer. In this scenario, it is not possible for any of the deterministic solutions to fulfill the fairness requirement, however, a randomized solution can be easily found to satisfy it. For instance, a product can be suggested from each group with the same likelihood of occurrence. We next summarize the main contributions of this paper.

1. We are the first to investigate the long-term fair submodular maximization problem, which presents a substantial challenge due to its exponential number of variables. As a result, it is difficult to solve using traditional linear programming (LP) solvers.
2. We develop a $(1 - 1/e)^2$-approximation algorithm that approximately satisfies the fairness constraints. Specifically, our algorithm ensures that the number of

selected items from group V_t is within the range of $[\lfloor \alpha_t \rfloor, \lceil \beta_t \rceil]$. Notably, if both α_t and β_t are integers, our solution strictly satisfies the fairness constraints in the original problem.
3. It is important to note that the previous algorithm requires optimizing a continuous approximation of the underlying submodular function, referred to as the multi-linear extension [6]. This is achieved by executing the continuous greedy algorithm, whose implementation is computationally expensive in practice. Our second contribution is the introduction of a fast greedy algorithm that achieves a degraded approximation ratio of $(1 - 1/e)^2/2$.
4. We present a $(1 - 1/e)$-approximation randomized algorithm. Our approach involves utilizing the ellipsoid method and incorporating an approximate separation oracle for the dual LP of the original problem, which has a polynomial number of variables and an exponential number of constraints. Unlike the deterministic solutions, our randomized approach provides three key benefits. Firstly, our solution does not depend on the assumption of non-overlapping groups. Secondly, our approach strictly satisfies all fairness constraints. Thirdly, we achieve the optimal approximation ratio of $1 - 1/e$.

2 Preliminaries and Problem Statement

A set V of n items is considered and there is a non-negative submodular utility function $f : 2^V \rightarrow \mathbb{R}_+$. The marginal utility of an item $e \in V$ on a set $S \subseteq V$ is denoted as $f(e \mid S)$, i.e., $f(e \mid S) = f(\{e\} \cup S) - f(S)$. The function f is considered submodular if, for any sets $X, Y \subseteq V$ with $X \subseteq Y$ and any item $e \in V \setminus Y$, the following inequality holds: $f(e \mid Y) \leq f(e \mid X)$. It is considered monotone if, for any set $X \subseteq V$ and any item $e \in V \setminus X$, it holds that $f(e \mid X) \geq 0$.

Assuming V is divided into m groups, $V_1, V_2, \ldots, V_m$, there is a specified lower and upper bound on the *expected* number of items from each group that must be included in a feasible solution. These bounds, referred to as $\alpha \in \mathbb{R}^m_{\geq 0}$ and $\beta \in \mathbb{R}^m_{\geq 0}$, represent group fairness constraints. In addition, there is a *hard* constraint b on the number of selected items. Let $\mathcal{F} = \{S \subseteq V \mid |S| \leq b\}$ denote the set of feasible selections. The goal of the fair submodular maximization problem (denoted as **P.0**) is to determine a distribution $x \in [0, 1]^{\mathcal{F}}$ over sets from $\mathcal{F}$ that maximizes the expected utility, while ensuring that the expected number of items selected from each group meets the fairness constraints. I.e.,

$$\textbf{P.0} \max_{x \in [0,1]^{\mathcal{F}}} \sum_{S \in \mathcal{F}} x_S f(S) \textbf{s.t.} \begin{cases} \alpha_t \leq \sum_{S \in \mathcal{F}} (x_S \cdot |S \cap V_t|) \leq \beta_t, \forall t \in [m]. \\ \sum_{S \in \mathcal{F}} x_S \leq 1. \end{cases}$$

Here each decision variable x_S represents the selection probability of $S \in \mathcal{F}$. This LP has a total of $2m + 1$ constraints, excluding the obvious constraints that specify that $x_S \geq 0$ for all $S \in \mathcal{F}$. Despite this, the number of variables in the LP problem is equal to the number of elements in $\mathcal{F}$, which can be exponential in n. As a result, conventional LP solvers are unable to solve this LP problem efficiently. The next lemma asserts that **P.0** is a problem that is NP-hard.

Lemma 1 *Problem P.0 is NP-hard.*

Proof We demonstrate this by reducing it to the classic cardinality constrained monotone submodular maximization problem, which we will describe below. □

Definition 1 The cardinality constrained monotone submodular maximization problem takes as input a collection of items V, a monotone submodular function $f : 2^V \rightarrow \mathbb{R}_+$, and a cardinality constraint b. The goal is to choose a subset of items $S \subseteq V$ that maximizes $f(S)$ while ensuring that $|S| \leq b$.

To show the reduction, we take an instance of the cardinality constrained monotone submodular maximization problem and create a corresponding instance of **P.0**. To do this, we consider only one group with no fairness constraints, meaning $V = V_1$, with $\alpha_1 = 0$ and $\beta_1 = |V|$. It can be easily verified that the optimal solution of this instance is a distribution over a set of solutions, each of which is an optimal solution to the instance of cardinality constrained monotone submodular maximization problem. Additionally, although **P.0** allows for randomized solutions, there exists at least one optimal solution that is a deterministic set. Specifically, every optimal solution of the cardinality constrained monotone submodular maximization problem must be an optimal solution to its corresponding instance of **P.0**. Hence, these two instances are equivalent. This concludes the proof of the reduction. □

3 Near Feasible Deterministic Algorithms

In this section, we present a deterministic algorithm for **P.0**. Here we assume that m groups do not overlap with each other. To begin, we introduce the multilinear extension of a monotone submodular function f. Given a vector $y \in [0, 1]^n$, let S_y be a random set where each item $i \in V$ is independently added to S_y with probability y_i. Then we let $F(y) = \mathbb{E}[f(S_y)] = \sum_{S \subseteq V} f(S) \prod_{i \in S} y_i \prod_{i \notin S}(1 - y_i)$.

We next introduce a new optimization problem **P.1**. The goal of **P.1** is to compute a vector $y \in [0, 1]^n$ that maximizes $F(y)$ such that $\alpha_t \leq \sum_{i \in V_t} y_i \leq \beta_t, \forall t \in [m]$ and $\sum_{t \in [m]} \sum_{i \in V_t} y_i \leq b$.

$$\textbf{P.1} \max_{y \in [0,1]^n} F(y) \textbf{s.t.} \begin{cases} \alpha_t \leq \sum_{i \in V_t} y_i \leq \beta_t, \forall t \in [m]. \\ \sum_{t \in [m]} \sum_{i \in V_t} y_i \leq b. \end{cases}$$

The following lemma establishes a connection between the optimal solution of problem **P.0** and that of problem **P.1**. This lemma serves as a crucial foundation for understanding the relationship between the two problems and allows for the development of a near optimal solution for **P.0** by solving **P.1**.

Lemma 2 *Let x^* denote the optimal solution of **P.0** and y^* denote the optimal solution of **P.1**, it holds that*

$$(1 - 1/e) \sum_{S \in \mathcal{F}} x^*_S f(S) \leq F(y^*). \tag{1}$$

Proof Let B be a polytope defined as the set of all vectors $y \in [0, 1]^n$ that meet the conditions in **P.1**, i.e.,

$$B = \{y \in [0, 1]^n \mid \alpha_t \leq \sum_{i \in V_t} y_i \leq \beta_t, \forall t \in [m]; \sum_{t \in [m]} \sum_{i \in V_t} y_i \leq b; 0 \leq y_i \leq 1, \forall i \in V\}. \tag{2}$$

Given the optimal solution x^* of **P.0**, we then introduce a vector $\hat{y} \in [0, 1]^n$ such that $\hat{y}_i = \sum_{S \in \mathcal{F}} x^*_S \cdot \mathbf{1}_{i \in S}$ where $\mathbf{1}_{i \in S} = 1$ if $i \in S$ and $\mathbf{1}_{i \in S} = 0$ otherwise. It is easy to verify that the value of $\hat{y}_i$ represents the probability of item i being selected according to the distribution defined by x^*. We next show that to prove this lemma, it suffices to prove that

$$\hat{y} \in B. \tag{3}$$

As established in [2], if f is monotone and submodular and $\hat{y}_i = \sum_{S \in \mathcal{F}} x^*_S \cdot \mathbf{1}_{i \in S}$, then $(1 - 1/e) \sum_{S \in \mathcal{F}} x^*_S f(S) \leq F(\hat{y})$. Here $1 - 1/e$ is also known as *correlation gap* of monotone submodular functions. Suppose (3) is true and y^* is the optimal solution of **P.1**, it holds that $F(\hat{y}) \leq F(y^*)$. Therefore, this lemma is a direct consequence of the observation that $(1 - 1/e) \sum_{S \in \mathcal{F}} x^*_S f(S) \leq F(\hat{y}) \leq F(y^*)$.

The rest of the proof is devoted to proving $\hat{y} \in B$. First, because x^* is a feasible solution of **P.0**, it holds that $\alpha_t \leq \sum_{S \in \mathcal{F}} (x^*_S \cdot |S \cap V_t|) \leq \beta_t, \forall t \in [m]$. It follows that $\alpha_t \leq \sum_{i \in V_t} \hat{y}_i \leq \beta_t, \forall t \in [m]$, this is because $\sum_{S \in \mathcal{F}} (x^*_S \cdot |S \cap V_t|) = \sum_{i \in V_t} \hat{y}_i$ represents the expected number of items being selected from group V_t according to the distribution defined by x^*. Second, because x^* is a feasible solution of **P.0**, the expected number of selected items according to the distribution defined by x^* is at most b. Hence, $\sum_{t \in [m]} \sum_{i \in V_t} \hat{y}_i \leq b$. Third, it is trivial to show that $0 \leq \hat{y}_i \leq 1, \forall i \in V$. This finishes the proof of $\hat{y} \in B$. □

3.1 Algorithm Design

We next present our algorithm. Initially, we use a continuous greedy algorithm to compute a fractional solution for **P.1**, which we then round to obtain an integral solution.

Continuous greedy algorithm. We first provide a detailed description of the continuous greedy algorithm (listed in Algorithm 1). The framework of this algorithm was first developed in [6] and we adapt it to find a fractional solution within polytope B (listed in (2)). Note that polytope B is not downward-closed, which presents unique challenges in our study. This algorithm maintains a fractional solution $y^l \in [0, 1]^n$,

Algorithm 1 Continuous Greedy Algorithm

1: Set $\delta = 9n^2, l = 0, y^0 = [0]^n$.
2: **while** $l < \delta$ **do**
3: For each $i \in V$, estimate $F(i \mid y^l)$
4: Find an optimal solution $z \in [0,1]^n$ to **P.A**
5: **P.A** *Maximize*$_y$ $\sum_{i \in V} y_i F(i \mid y^l)$ **s.t.** $y \in B$.
6: $y^{l+1} = y^l + z$, increment $l = l + 1$
7: $y' \leftarrow y^\delta$
8: **return** y'

starting with $y^0 = (0, 0, \ldots, 0)$. In each round l, it computes the marginal utility of each item $i \in V$ on top of y^l with respect to F as follows,

$$F(i \mid y^l) = F(\mathbf{e}_i \vee y^l) - F(y^l). \tag{4}$$

where $\mathbf{e}_i \in \{0, 1\}^n$ is the vector with 1 in the i-th coordinate and 0 elsewhere; $\mathbf{e}_i \vee y^l$ denotes the element-wise maximum of two vectors $\mathbf{e}_i$ and y^l.

Then we solve the following linear programming problem **P.A** which assigns a weight $F(i \mid y^l)$ to each item i and seeks the maximum weighted vector in B.

$$\textbf{P.A} \textit{Maximize}_y \sum_{i \in V} y_i F(i \mid y^l) \textbf{s.t.} y \in B.$$

After solving **P.A** at round l and obtaining an optimal solution $z \in [0, 1]^n$, we update the fractional solution as follows: $y^{l+1} = y^l + z$. After δ rounds where $\delta = 9n^2$, y^δ is returned as the final solution y'.

Rounding. We next employ pipage rounding [1], a simple deterministic procedure of rounding of linear relaxations, to round y' to an integral solution. This algorithm is composed of three phases.

- Phase 1: For each $t \in [m]$, repeatedly perform the following until V_t has no more than one non-integral coordinate: Choose any two fractional coordinates i, j such that $i, j \in V_t$. Calculate $\theta_1 = \min\{1 - y'_i, y'_j\}$ and $\theta_2 = \min\{y'_i, 1 - y'_j\}$. Create two vectors, $y^a = y' + \theta_1(\mathbf{e}_i - \mathbf{e}_j)$ and $y^b = y' + \theta_2(\mathbf{e}_j - \mathbf{e}_i)$. If $F(y^a) \geq F(y^b)$, set $y \leftarrow y_a$, otherwise set $y \leftarrow y^b$.
- Phase 2: Assume $y_1, \ldots, y_k$ are the remaining fractional coordinates. Repeat the same procedure as in the first phrase until y has at most one non-integral coordinate.
- Phase 3: Let i denote the last non-integral coordinate, if any. Set $y_i = 1$. Output $A \subseteq V$ whose coordinate in y is 1.

Note that a similar framework has been utilized to tackle the fair submodular maximization problem in a deterministic setting [7]. This problem aims to identify a fixed set of items that optimize a submodular function while fulfilling group fairness constraints. Their approach shares similarities with ours in the rounding stage, but does not require the third phase. This is because in their setting, both α_t and β_t are integers, which allows them to ensure that no non-integral coordinates exist after the first two rounding phases.

3.2 Performance Analysis

Recall that x^* denotes the optimal solution of **P.0**, let $OPT = \sum_{S\in\mathcal{F}} x^*_S f(S)$ denote the utility of the optimal solution. The following theorem states that A, the solution set returned from our algorithm, is a *near* feasible solution of **P.0** and has a utility of at least $(1-1/e)^2 OPT$.

Theorem 1 *Let A be the set returned by our algorithm and OPT be the utility of the optimal solution of* ***P.0****. It follows that:*

$$f(A) \geq (1-1/e)^2 OPT. \tag{5}$$

Moreover, A always satisfies the cardinality constraint and nearly satisfies the fairness constraints of ***P.0****, i.e., $|A| \leq b$ and $\lfloor \alpha_t \rfloor \leq |A \cap V_t| \leq \lceil \beta_t \rceil, \forall t \in [m]$.*

Proof We first prove that $|A| \leq b$ always holds. Observe that the fractional solution y' found by the continuous greedy algorithm belongs to $\mathcal{B}$, hence, $\sum_{i\in V} y'_i \leq b$. Moreover, phases 1 and 2 in the rounding stage do not change this value, and phase 3 rounds the last non-integral coordinate, if any, to one. It follows that $|A| \leq \lceil \sum_{i\in V} y'_i \rceil \leq b$ where the second inequality is by the observations that $\sum_{i\in V} y'_i \leq b$ and b is an integer.

We next prove that A nearly satisfies the fairness constraints of **P.0**, i.e., $\lfloor \alpha_t \rfloor \leq |A \cap V_t| \leq \lceil \beta_t \rceil, \forall t \in [m]$. Because $y' \in \mathcal{B}$, it holds that $\alpha_t \leq \sum_{i\in V_t} y'_i \leq \beta_t, \forall t \in [m]$. Observe that phase 1 does not change this value, phases 2 and 3 round at most one fractional coordinate from each group to a binary value. Hence, $\lfloor \sum_{i\in V_t} y'_i \rfloor \leq |A \cap V_t| \leq \lceil \sum_{i\in V_t} y'_i \rceil, \forall t \in [m]$. This, together with $\alpha_t \leq \sum_{i\in V_t} y'_i \leq \beta_t, \forall t \in [m]$, implies that $\lfloor \alpha_t \rfloor \leq |A \cap V_t| \leq \lceil \beta_t \rceil, \forall t \in [m]$.

At last, we prove the approximation ratio of A. Recall that y^* denotes the optimal solution of **P.1**, [6] has proved that if f is monotone and submodular, then the fractional solution y' returned from the continuous greedy algorithm has a utility of at least $(1-1/e)F(y^*)$, i.e., $F(y') \geq (1-1/e)F(y^*)$. This, together with Lemma 2, implies that $F(y') \geq (1-1/e)^2 \sum_{S\in\mathcal{F}} x^*_S f(S) = (1-1/e)^2 OPT$. To prove $f(A) \geq (1-1/e)^2 OPT$, it suffices to show that $f(A) \geq F(y')$. We next prove this inequality. Observe that in phases 1 and 2 of the rounding stage, we perform pipage rounding to round y' to a vector y that contains at most one non-integral coordinate. According to [6], pipage rounding does not decrease the expected utility of y', that is, $F(y) \geq F(y')$. In phase 3, we round the last non-integral coordinate in y, if any, to one. This operation does not decrease the expected utility of y by the assumption that f is monotone. Hence, $F(y) \geq F(y')$ still holds. Recall that y is the indicator vector of A, hence, $f(A) = F(y)$. Therefore, $f(A) = F(y) \geq F(y')$. □

Remark 1 It follows immediately from the preceding theorem that if α_t and β_t are both integers for all $t \in [m]$, then our solution strictly satisfies all fairness constraints of problem **P.0**.

3.3 A Fast Greedy Algorithm

Our prior algorithm involves solving a multi-linear relaxation problem, which can be slow and computationally expensive, particularly for large scale problems. In this section, we introduce a simple greedy algorithm that offers a significant increase in speed but with a trade-off in the form of a decreased approximation ratio.

Even though **P.0** permits the use of randomized solutions, Theorem 1 shows that a deterministic solution is sufficient for obtaining a constant-factor approximation for **P.0**. We next present a simple greedy algorithm that effectively finds a near optimal deterministic solution, which in turn results in a constant-factor approximation for the problem **P.0**. To this end we introduce a new optimization problem **P.2**, a deterministic version of **P.0** (with relaxed fairness constraints).

$$\textbf{P.2} \max_{S \in \mathcal{F}} f(S) \textbf{s.t.} \lfloor \alpha_t \rfloor \leq |S \cap V_t| \leq \lceil \beta_t \rceil, \forall t \in [m].$$

Note that in **P.2** we use $\lfloor \alpha_t \rfloor$ and $\lceil \beta_t \rceil$ as lower and upper bounds, hence a feasible solution of **P.2** is a near feasible solution of the original problem **P.0**. The following lemma states that the optimal solution of **P.2** attains a $(1 - 1/e)^2$ approximation of the problem **P.0**.

Lemma 3 *Let A^{P2} denote the optimal solution of* **P.2**, *it holds that $f(A^{P2}) \geq (1 - 1/e)^2 OPT$ where OPT is the optimal solution of* **P.0**.

Proof Recall that in Theorem 1, we show that $f(A) \geq (1 - 1/e)^2 OPT$ where A satisfies all constraints in **P.2**. Because A^{P2} is the optimal solution of **P.2**, we have $f(A^{P2}) \geq f(A) \geq (1 - 1/e)^2 OPT$. □

We next present a simple greedy algorithm to attain a $1/2$ approximation of **P.2**. First, we present **P.3**, a relaxed problem of **P.2**.

$$\textbf{P.3} \max_{S \subseteq V} f(S) \textbf{s.t.} \begin{cases} |S \cap V_t| \leq \lceil \beta_t \rceil, \forall t \in [m]. \\ \sum_{t \in [m]} \max\{\lfloor \alpha_t \rfloor, |S \cap V_t|\} \leq b. \end{cases}$$

It is easy to verify that any feasible solution of **P.2** must be a feasible solution of **P.3**. Hence, $f(A^{P2}) \leq f(A^{P3})$ where A^{P3} is the optimal solution of **P.3**. It has been shown that the constraints listed in **P.3** constitute a matroid [13]. Hence, **P.3** is a classic submodular maximization problem subject to a matroid constraint. A simple greedy algorithm guarantees a $1/2$ approximation of **P.3** [17]. This algorithm works by iteratively adding items to the solution set such that at each step, the marginal increase in the objective value is maximized, and the matroid constraint is satisfied, and it terminates when the current solution set can not be expanded. Let A^g denote the solution returned from the greedy algorithm, it holds that

$$f(A^g) \geq (1/2)f(A^{P3}) \geq (1/2)f(A^{P2}). \tag{6}$$

Moreover, it is easy to verify that A^g must be a feasible solution of **P.2**.

Theorem 2 *Let A^g denote the solution of returned from the greedy algorithm, it holds that $f(A^g) \geq \frac{(1-1/e)^2}{2} \cdot OPT$. Moreover, A^g always satisfies the cardinality constraint and nearly satisfies the fairness constraints of **P.0**, i.e., $|A^g| \leq b$ and $\lfloor \alpha_t \rfloor \leq |A^g \cap V_t| \leq \lceil \beta_t \rceil, \forall t \in [m]$.*

Proof The proof of the first part of this theorem stems from inequality (6) and Lemma 3. The second part of this theorem is because A^g is a feasible solution to problem **P.2**. □

4 A Feasible $(1 - 1/e)$-Approximation Randomized Algorithm

We now present a randomized algorithm for **P.0**. In contrast to the results presented in the previous section, our randomized solution offers three advantages: (1) our solution does not rely on the assumption of non-overlapping groups, (2) our solution satisfies all fairness constraints in a *strict* sense, and (3) we achieve the optimal approximation ratio of $1 - 1/e$.

As previously stated, **P.0** has a number of variables equal to the number of elements in $\mathcal{F}$, which can become extremely large when n is significant. This means that standard LP solvers are unable to handle this LP problem effectively. To tackle this issue, we resort to its corresponding dual problem (**Dual of P.0**) and utilize the ellipsoid algorithm [15] to solve it. In the dual problem, we assign two "weights" $z_t \in \mathbb{R}_{\geq 0}$ and $u_t \in \mathbb{R}_{\geq 0}$ to each group V_t and there is an additional variable $w \in \mathbb{R}_{\geq 0}$.

Dual of P.0 $\min_{z \in \mathbb{R}^m_{\geq 0}, u \in \mathbb{R}^m_{\geq 0}, w \in \mathbb{R}_{\geq 0}} \sum_{t \in [m]} (\beta_t u_t - \alpha_t z_t) + w$
s.t. $w \geq f(S) + \sum_{t \in [m]} |S \cap V_t| \cdot (z_t - u_t), \forall S \in \mathcal{F}.$

The ellipsoid method determines the emptiness of a non-degenerate convex set C, such as the feasible region of **Dual of P.0**. It defines an ellipsoid containing C and iteratively checks if the center is in C. If feasible, it explores smaller objectives. If not, it employs a (approx.) separation oracle, constructing smaller ellipsoids. This geometric process continues until a feasible solution is found or C is considered empty. The method requires a poly-time (approx.) separation oracle and has polynomial iterations for linear problems.

In the context of our problem, we approximately solve the SUBMAX problem to check the feasibility of the current solution and act as the separation oracle.

Definition 2 Given a utility function f, a cardinality constraint b, and two vectors $z \in \mathbb{R}^m_{\geq 0}$ and $u \in \mathbb{R}^m_{\geq 0}$, SUBMAX$(z, u, b)$ aims to $\max_{S \in \mathcal{F}}(f(S) + \sum_{t \in [m]} |S \cap V_t| \cdot (z_t - u_t))$.

SUBMAX(z, u, b) asks for a set S of size at most b such that $f(S) + \sum_{t \in [m]} |S \cap V_t| \cdot (z_t - u_t)$ is maximized. Observe that f is non-negative monotone and submodular; and $\sum_{t \in [m]} |S \cap V_t| \cdot (z_t - u_t)$ is a modular function in terms of S, hence, SUBMAX(z, u, b) is a classic problem of maximizing the summation of a non-negative monotone submodular and a modular function under cardinality constraints. Reference [19] developed a randomized algorithm that finds a set A such that for every $S \in \mathcal{F}$, it holds that

$$f(A) + \sum_{t \in [m]} |A \cap V_t| \cdot (z_t - u_t) \geq (1 - 1/e) f(S) + \sum_{t \in [m]} |S \cap V_t| \cdot (z_t - u_t). \quad (7)$$

Now we are ready to present the main theorem of this section.

Theorem 3 *There exists a polynomial time $(1 - 1/e)$-approximation algorithm (with additive error ϵ) for* ***P.0****.*

The rest of this section is devoted to proving Theorem 3, that is, we present a polynomial $(1 - 1/e)$-approximation algorithm for **P.0**. Let $C(L)$ denote the set of $(z \in \mathbb{R}^m_{\geq 0}, u \in \mathbb{R}^m_{\geq 0}, w \in \mathbb{R}_{\geq 0})$ satisfying that

$$\sum_{t \in [m]} (\beta_t u_t - \alpha_t z_t) + w \leq L \text{ and } w \geq f(S) + \sum_{t \in [m]} |S \cap V_t| \cdot (z_t - u_t), \forall S \in \mathcal{F}.$$

We use binary search to determine the smallest value of L for which $C(L)$ is not empty. Given a specific value of L, we first check the inequality $\sum_{t \in [m]}(\beta_t u_t - \alpha_t z_t) + w \leq L$. Then, we run algorithm from [19] (labeled as $\mathcal{A}$) to solve SUBMAX(z, u, b). Let A be the solution set returned from $\mathcal{A}$.

- If $f(A) + \sum_{t \in [m]} |A \cap V_t| \cdot (z_t - u_t) \leq w$, then $C(L)$ is marked as non-empty. In this case, we try a smaller L.
- If $f(A) + \sum_{t \in [m]} |A \cap V_t| \cdot (z_t - u_t) > w$, then $(z, w) \notin C(L)$ and A is a separating hyperplane. We identify a reduced-size ellipsoid whose center complies with the given constraint. This process continues until a feasible solution in $C(L)$ is found (in this case, we try a smaller L) or the volume of the bounding ellipsoid is so small that $C(L)$ is considered empty (in this case, it is evident that reaching the current objective is not achievable and therefore, we will attempt a larger value for L).

To learn about the specific steps involved in running ellipsoid using separation oracles and achieving (multiplicative and additive) approximate guarantees, we suggest referring to Chap. 2 of [5]. Assume L^* is the minimum value of L for which the algorithm determines that $C(L)$ is non-empty. Hence, there exists a (z^*, u^*, w^*) such that

$$\sum_{t\in[m]}(\beta_t u_t^* - \alpha_t z_t^*) + w^* \leq L^* \tag{8}$$

and

$$f(A) + \sum_{t\in[m]} |A \cap V_t| \cdot (z_t^* - u_t^*) \leq w^* \tag{9}$$

where A is the output obtained from $\mathcal{A}$ after solving SUBMAX(z^*, u^*, b). Let $\mu = 1 - 1/e$, it follows that $\forall S \in \mathcal{F}$,

$$\begin{aligned} f(S) + \sum_{t\in[m]} |S \cap V_t| \cdot (u_t^* - z_t^*)/\mu &\leq (f(A) + \sum_{t\in[m]} |A \cap V_t| \cdot (u_t^* - z_t^*))/\mu \\ &\leq w^*/\mu \end{aligned} \tag{10}$$

where the first inequality is by (7) and the second inequality is by inequality (9). In addition, inequality (8) implies that

$$\sum_{t\in[m]}(\beta_t u_t^* - \alpha_t z_t^*)/\mu + w^*/\mu \leq L^*/\mu. \tag{11}$$

In Eq. (10) implies $(z^*/\mu, u^*/\mu, w^*/\mu)$ is a feasible solution of **Dual of P.0**. This, together with in Eq. (11), implies that the optimal solution of **Dual of P.0** and thus the optimal solution of **P.0** (by strong duality) is upper bounded by $\frac{1}{\mu} \cdot L^*$. By solving **P.0** with a value of L^*, we attain a μ-approximation solution for the original problem **P.0**. Here, we explain how to compute such a solution using only feasible solution sets corresponding to the separating hyperplanes found by the separation oracle. Assume $L^* - \epsilon$ is the largest value of L for which the algorithm determines that $C(L)$ is empty, where ϵ is decided by the precision of our algorithm. Let $\mathcal{F}'$ denote the subset of $\mathcal{F}$ for which the dual constraint is violated during the execution of the ellipsoid algorithm on $C(L^* - \epsilon)$. Then, $|\mathcal{F}'|$ is polynomial. We consider the following polynomial sized **Dual of P.0** (labeled as **Poly-sized Dual of P.0**), using separating hyperplanes from $\mathcal{F}'$.

Poly-sized Dual of P.0 $\min_{z\in\mathbb{R}^m_{\geq 0}, u\in\mathbb{R}^m_{\geq 0}, w\in\mathbb{R}_{\geq 0}} \sum_{t\in[m]}(\beta_t u_t - \alpha_t z_t) + w$
s.t. $w \geq f(S) + \sum_{t\in[m]} |S \cap V_t| \cdot (z_t - u_t), \forall S \in \mathcal{F}'$.

Because $C(L^* - \epsilon)$ is empty, the optimal solution to **Poly-sized Dual of P.0** is at least $L^* - \epsilon$. Hence, the value of the dual of **Poly-sized Dual of P.0**, which is listed in **Poly-sized P.0**, is at least $L^* - \epsilon$. Note that **Poly-sized P.0** is of polynomial size.

$$\textbf{Poly-sized P.0} \max_{x\in[0,1]^{\mathcal{F}'}} \sum_{S\in\mathcal{F}'} x_S f(S)$$
$$\textbf{s.t.} \begin{cases} \alpha_t \leq \sum_{S\in\mathcal{F}} (x_S \cdot |S \cap V_t|) \leq \beta_t, \forall t \in [m]. \\ \sum_{S\in\mathcal{F}'} x_S \leq 1. \end{cases}$$

Recall that the optimal solution of **P.0** is upper bounded by $\frac{1}{\mu} \cdot L^*$, obtaining an optimal solution from **Poly-sized P.0** provides a μ-approximation for **P.0** (with additive error ϵ), where $\mu = 1 - 1/e$.

References

1. Ageev, A.A., Sviridenko, M.I.: Pipage rounding: a new method of constructing algorithms with proven performance guarantee. J. Comb. Optim. **8**, 307–328 (2004)
2. Agrawal, S., Ding, Y., Saberi, A., Ye, Y.: Correlation robust stochastic optimization. In: SODA, pp. 1087–1096. SIAM (2010)
3. Bera, S., Chakrabarty, D., Flores, N., Negahbani, M.: Fair algorithms for clustering. NIPS **32** (2019)
4. Biddle, D.: Adverse impact and test validation: a practitioner's guide to valid and defensible employment testing. Routledge (2017)
5. Bubeck, S., et al.: Convex optimization: algorithms and complexity. Found. Trends® Mach. Learn. **8**(3-4), 231–357 (2015)
6. Calinescu, G., Chekuri, C., Pál, M., Vondrák, J.: Maximizing a submodular set function subject to a matroid constraint. In: IPCO, pp. 182–196. Springer, Berlin (2007)
7. Celis, L.E., Huang, L., Vishnoi, N.K.: Multiwinner voting with fairness constraints. In: IJCAI, pp. 144–151 (2018)
8. Chierichetti, F., Kumar, R., Lattanzi, S., Vassilvtiskii, S.: Matroids, matchings, and fairness. In: AISTATS, pp. 2212–2220. PMLR (2019)
9. Das, A., Kempe, D.: Algorithms for subset selection in linear regression. In: Proceedings of the Fortieth Annual ACM Symposium on Theory of Computing, pp. 45–54 (2008)
10. Dueck, D., Frey, B.J.: Non-metric affinity propagation for unsupervised image categorization. In: 2007 IEEE 11th International Conference on Computer Vision, pp. 1–8. IEEE (2007)
11. Dwork, C., Hardt, M., Pitassi, T., Reingold, O., Zemel, R.: Fairness through awareness. In: Proceedings of the 3rd Innovations in Theoretical Computer Science Conference (2012)
12. El-Arini, K., Guestrin, C.: Beyond keyword search: discovering relevant scientific literature. In: KDD, pp. 439–447 (2011)
13. El Halabi, M., Mitrović, S., Norouzi-Fard, A., Tardos, J., Tarnawski, J.M.: Fairness in streaming submodular maximization: algorithms and hardness. NIPS **33**, 13609–13622 (2020)
14. Golovin, D., Krause, A.: Adaptive submodularity: theory and applications in active learning and stochastic optimization. J. Artif. Intell. Res. **42**, 427–486 (2011)
15. Grötschel, M., Lovász, L., Schrijver, A.: The ellipsoid method and its consequences in combinatorial optimization. Combinatorica **1**(2), 169–197 (1981)
16. Monroe, B.L.: Fully proportional representation. Am. Polit. Sci. Rev. **89**(4), 925–940 (1995)
17. Nemhauser, G.L., Wolsey, L.A., Fisher, M.L.: An analysis of approximations for maximizing submodular set functions-i. Math. Program. **14**(1), 265–294 (1978)
18. Sipos, R., Swaminathan, A., Shivaswamy, P., Joachims, T.: Temporal corpus summarization using submodular word coverage. In: Proceedings of the 21st ACM International Conference on Information and Knowledge Management, pp. 754–763 (2012)
19. Sviridenko, M., Vondrák, J., Ward, J.: Optimal approximation for submodular and supermodular optimization with bounded curvature. Math. Oper. Res. **42**(4), 1197–1218 (2017)

20. Tang, S.: Beyond pointwise submodularity: non-monotone adaptive submodular maximization in linear time. Theoret. Comput. Sci. **850**, 249–261 (2021)
21. Tang, S.: Beyond pointwise submodularity: non-monotone adaptive submodular maximization subject to knapsack and k-system constraints. Theoret. Comput. Sci. **936**, 139–147 (2022)
22. Tang, S., Yuan, J.: Influence maximization with partial feedback. Oper. Res. Lett. **48**(1), 24–28 (2020)
23. Tang, S., Yuan, J.: Group equility in adaptive submodular maximization. INFORMS J. Comput. (2023)
24. Tang, S., Yuan, J.: Optimal sampling gaps for adaptive submodular maximization. In: AAAI (2022)
25. Yuan, J., Tang, S.: Group fairness in non-monotone submodular maximization. J. Comb. Optim. **45**(3), 88 (2023)

On Syntactical Graphs-of-Words

Nabil Moncef Boukhatem, Davide Buscaldi, and Leo Liberti

Abstract A graph-of-words is a graph representation of natural language text based on proximity in the linear text reading order: the vertices are the words, and edges are induced by k left and right neighbours of the words. Vertices representing same or similar words are then contracted. We propose graphs-of-words where edges are instead induced on paths in the syntax trees (we investigate both dependency and constituency trees). We discuss some properties, advantages, and disadvantages of classic and new graphs-of-words on texts extracted from literature, as well as from a technical Q&A database.

1 Introduction

Natural language is human-specific, ambiguous, and often ungrammatical; its understanding is usually subjected to context knowledge. It is opposed to formal language, which is computer-specific, unambiguous, and must be grammatically perfect to be meaningful: its pragmatics are formally defined by the effect it has on an electronic or mechanical system. In this paper we use formal language constructs to instruct computers to deal with natural language text. More precisely, we focus on a very specific and well-known task in Natural Language Processing (NLP), i.e. that of *key-*

N. M. Boukhatem · L. Liberti (✉)
LIX CNRS, Ecole Polytechnique, 91128 Palaiseau, France
e-mail: leo.liberti@polytechnique.edu

N. M. Boukhatem
OneTeam, Paris, France

Present Address:
Lundimatin, Montpellier, France
e-mail: nboukhatem@oneteam.fr; moncef-nabil.boukhatem@lundimatin.fr

D. Buscaldi
LIPN CNRS, Université de Paris-Nord, Villetaneuse, France
e-mail: buscaldi@lipn.univ-paris13.fr

A. Brieden et al. (eds.), *Graphs and Combinatorial Optimization: from Theory to Applications*, AIRO Springer Series 13,
https://doi.org/10.1007/978-3-031-46826-1_14

word extraction: given a text in natural language, output K keywords that a human would find the most pertinent for the text. This obviously poses the issue of empirical verification: since different humans would have different preferences, how do we determine what keywords are "best"? In this paper we resort to *ground truths* put together by a restricted number of humans (see Sect. 3).

The most established method for extracting keywords from natural language text is probably based on ranking functions (see e.g. [14]) based on frequency of words in documents with respect to a set of documents called *corpus* [11]. The main application is automatic document indexing or summarization [12].

This paper replaces the concept of word frequency in documents with that of vertex degrees in graphs that represent the text. Methodologically speaking, the main contribution is a comparison between different graph representations of text. One of the graph representations we consider is derived from constituency syntax trees (see Sect. 1.3, which, to the best of our knowledge, has never been previously considered for this purpose).

1.1 Ranking Functions for Text

The earliest cornerstone of information retrieval in text is perhaps the TF-IDF ranking function. It consists of the product of two other functions: Term Frequency (TF) and Inverse Document Frequency (IDF) [15]. We shall limit our introduction to the functions we actually used in our computational experiments. In the following formulæ, we let C be a corpus (i.e., a set) of text documents, D be a document in C, and t be a term (i.e., a word) in D. Then:

$$\mathsf{tf}(t, D) = |(v \in D \mid v = t)| \tag{1}$$

$$\mathsf{TF}(t, D) = 1 + \ln(1 + \max(0, \ln(\mathsf{tf}(t, D)))) \tag{2}$$

$$\mathsf{IDF}(t, C) = \frac{\ln(|C| + 1)}{\sum_{D \in C} \mathsf{tf}(t, D)} \tag{3}$$

are the basic building blocks for two well-known ranking functions. These are:

$$\mathsf{TFIDF}(t, D, C) = \frac{\mathsf{TF}(t, D)\,\mathsf{IDF}(t, C)}{1 - b + \left(\frac{b|D|}{\sum_{P \in C} |P|/|C|}\right)} \tag{4}$$

$$\mathsf{BM25}(t, D, C) = \frac{(k_1 + 1)\mathsf{tf}(t, D)\,\mathsf{IDF}(t, C)}{k_1\left(1 - b + \left(\frac{b|D|}{\sum_{P \in C} |P|/|C|}\right)\right) + \mathsf{tf}(t, D)}, \tag{5}$$

where $b = 0.5$ and $k_1 = 1.2$. We aim at replacing TF with the weighted degree $\mathsf{tw}(t, D)$ of a word vertex t in a graph $G(D)$ representing a document, namely:

$$\mathsf{tw}(t, D) = \sum_{v \in N_{G(D)}(t)} d_{tv} \tag{6}$$

$$\mathsf{TWIDF}(t, D, C) = \frac{\mathsf{tw}(t, D)\,\mathsf{IDF}(t, C)}{1 - b + b(\frac{|D|}{\sum_{D \in C} |P|/|C|})}, \tag{7}$$

where d_{uv} is the weight of the edge $\{u, v\}$ in the graph $G(D)$, and $b = 0.75$. The weights of the constants is taken from [16]. For unweighted graphs $G(D)$ we have $d_{uv} = 1$ for all edges $\{u, v\}$.

1.2 Graph-of-Words

Graphs can be used to summarize and extract keywords from a text in natural language [13]. In general, these graphs encode syntactical and sometimes semantic information on the edges, that represent relations on the words inferred from the text. Here we look at a purely syntactical construction proposed in Rousseau's Ph.D thesis [16] under the name *graph-of-words*.

In a graph-of-words (gow), the vertices are labeled by the words. The edges incident to each node are induced by the proximity of the words that are left and right of the node word in the linear text reading order. For example, in the sentence "Computers are close to understanding natural language", the words "are" and "to" are 1-proximal to "close", and the words "computers" and "understanding" are 2-proximal to "close". In a gow with proximity parameter 2, the node labelled by "close" would be adjacent to the vertices labelled by "computers", "are", "to", "understanding".

Note that, if a word occurs more than once in a text, this construction creates separate vertices referring to each occurrence. Moreover, the resulting graph would be a simple chain of embedded cliques, where almost every vertex has the same degree. This motivates a last contraction step in the construction of gows: if two or more vertices represent different occurrences of the same word, they are contracted to a single node. This last step is sometimes interpreted more broadly, for example by contracting vertices having same *lemmatized* word (i.e. the stem of the word without the desinences). An important pre-processing step to a useful gow is the removal of *stop-words*: words that are very frequent in most texts, but do not carry keyword-status information. Typically, stop-words are articles, auxiliary verbs or particles, prepositions, conjunctions, common adverbs, and so on. An example of a gow is given in Fig. 1. We note that gows of proximity k have at least $2k$ adjacencies.

1.3 Syntax Trees

In the framework of formal languages, *syntax trees* are the trace of a parsing algorithm for the sentences of the language. They also provide the mechanism by which

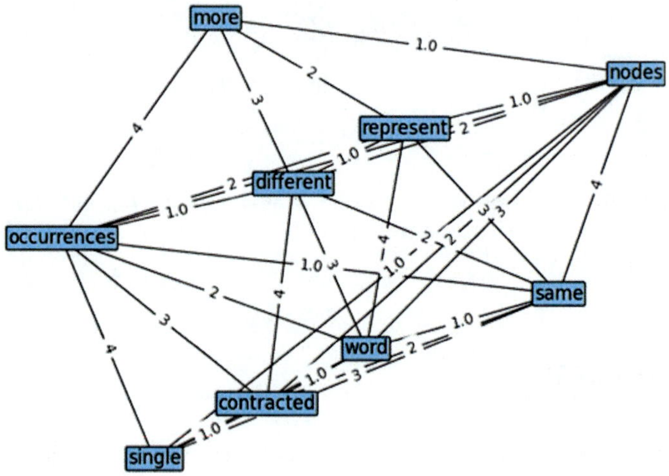

Fig. 1 A graph-of-word with proximity 2 of the sentence "if two or more nodes represent different occurrences of the same word, they are contracted to a single node". Edges are weighted by the text distance between the two word vertices, but this weighting is not essential. The node with largest degree is labelled by the word "nodes" (in the above graph, the two nodes corresponding to "nodes" and "node" were contracted)

computers assign semantics to high-level programs, or, in other words, execute code [6, 10]. Parsing algorithms use a *formal grammar* in order to drive a recursive analysis of a formal language sentence. The grammar consists of a set of rules of the form

$$\mathsf{tag} \longrightarrow \mathsf{comp}_1^1 \ldots \mathsf{comp}_{n_1}^1 \mid \ldots \mid \mathsf{comp}_1^h \ldots \mathsf{comp}_{n_k}^h,$$

which requires that a phrase tag be decomposed in one of h ways, each of which consists of a certain number of components, which can themselves be phrase tags or words. The grammar includes rules for each of the component tags down to the words, which are part of a given vocabulary. Each sentence input is assigned an initial tag, e.g. S for "sentence". The parser resolves tags recursively in terms of the component tags prescribed by the grammar rules, for as long as there are relevant rules that apply. In so doing, the parser produces a syntax tree. If the parser stops before all tags are resolved into constant words, the sentence does not conform to the grammar rules (this is how interpreters and compilers flag syntax errors). Otherwise, the recursive parsing process can also assign executable machine code to each of the constant words (which may be loops, tests, assignments), and then compose the code into an executable program (this is how interpreters and compilers turn a high-level language program into a set of actions performed by the CPU).

Noam Chomsky is credited with the popularization of syntax trees applied to natural languages [1], where the sentence tag S is usually mapped to the decomposition NP VP: i.e., a sentence corresponds to a noun phrase and a verb phrase. These two tags are then recursively decomposed until the words are reached. Since natural lan-

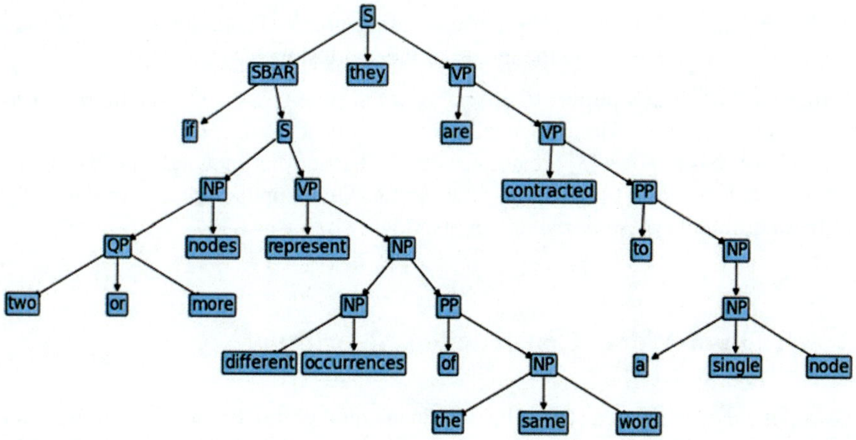

Fig. 2 The constituency tree for the same sentence as in Fig. 1. The tags are: S (sentence), SBAR (subordinate sentence), NP (noun phrase), VP (verb phrase), QP (quantificational phrase), PP (propositional phrase)

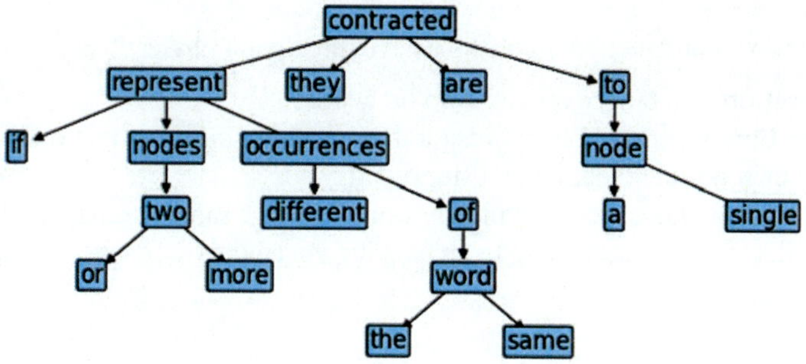

Fig. 3 The dependency tree for the same sentence as in Fig. 1. The arcs are usually labeled by the dependency tag of a child node to its parent node, not shown here because they are not used in this paper

guage is not formal, in general there may be many possible recursive decompositions, all leading to a different meaning, without an obvious way to choose between them. Chomsky's trees are called *constituency trees* (see Fig. 2 for an example).

Dependency trees are different types of trees originally introduced to linguistics by Louis Tesnière. The root of the tree is the main verb of the sentence, which has subject and main complement as child nodes. Each noun node has articles, adjectives, adverbs as child nodes (see Fig. 3 for an example).

Our interest in syntax trees is that they provide a binary relation on words alternative to linear text order proximity. For dependency trees, this order is natural. For constituency trees, the words in a sentence appear as leaf nodes. In both cases, since (undirected) trees are connected, each word is adjacent to any other word by means of the shortest path between the corresponding nodes. This allows us to define a natural edge weight equal to the length of the shortest path.

Contributions of this paper. In this paper we present gows based on different syntactical relations: k-proximity, dependency, constituency. While k-proximity [17] and dependency-based gows [3] are not new, to the best of our knowledge, constituency trees were never used to construct gows so far. We computationally evaluate gows of these different types on several counts.

2 Graph-of-Words Construction Algorithms

By a *sentence* we mean a string that a human could correctly transform into a valid syntax tree. A *phrase* is a sub-string of a sentence, which appears as a sub-tree of the sentence's syntax tree. Sentences are also assumed to be equivalent to lists of *tokens*, where each token can be either a word or a punctuation symbol. Notation-wise, for a sentence s we let s_i be the ith token of s for every $i \leq |s|$, which is the number of tokens of s.

All our gow construction algorithms have three main phases:

1. **generation** of a binary relationship on words;
2. **projection** over important words (and removal of non-important ones: typically these includes punctuation and stop-words);
3. **contraction** of like words (typically words with the same lemmatization, or with a similar meaning according to an existing vocabulary or encyclopedia [2]).

2.1 Proximity Gows

In proximity gows the two phases (generation, projection) may be carried out in either order, but changing the order yields different weights (usefulness of edge weights in proximity gow is doubtful, though [16]). For a string of n tokens, the **generation** phase is as follows. Initially, $V = \{s_1, \ldots, s_n\}$ and E is empty. Then we add edges $\{s_i, s_{i-h}\}$ and $\{s_i, s_{i+h}\}$ for all $1 \leq h \leq k$ and for all $h < i < n - h$.

The **projection** phase, if carried out before generation, simply removes the tokens deemed unimportant from the sentence s. The new list of tokens s' is then subjected to the generation phase. Otherwise projection re-arranges edges incident to removed token vertices: we iteratively replace pairs of edges $(\{v, u\}, \{u, w\})$ incident to a removed vertex u by means of an edge $\{v, w\}$ with weight $d_{vw} = d_{vu} + d_{uw}$. Note that the removal process may add an edge $\{v, w\}$ involving a removed vertex: this edge will be part of a replaced pairs later in the iteration.

Proposition 1 *Let $G = (V, E)$ be the the k-proximity graph obtained from the sentence $s = (s_1, \ldots, s_n)$ by performing generation first, then projection; and $H = (U, F)$ be obtained by projection then generation. We have $G = H$.*

Proof We have $V = U$ because projection removes the same vertices whether carried out before or after generation. Let us now consider an edge $\{u, v\} \in E$, where $u = s_i$ and $v = s_j$ for some $i < j$. If $j - i \leq k$ in the original sentence s then projection either leaves $j - i$ invariant or makes it smaller, so $\{u, v\} \in F$. Assume now that $j - i = k + 1$. This means that there is an index h with $i < h < j$ such that s_h is a removed node. Then, after generation, there must be an edge pair $(\{s_i, s_h\}, \{s_h, s_j\})$ in the graph that is replaced by a single edge $\{s_i, s_j\}$: obviously, since s_h is removed first in H, this edge is also in F. By induction, the same holds for any value of $j - i > k$. The argument showing that edges in F must also be in E is similar. □

Given a weighted graph $G = (V, E, d)$ where V is a set of tokens of a string s, the **contraction** in G of a subset $R \subset V$ s.t. $|R| \geq 2$ is as follows: (i) a representative $r \in R$ is chosen; (ii) in all edges $\{v, u\} \in E$ with $v \notin R$ and $u \in R$ the symbol u is replaced by r, with $d_{vr} = d_{vu} + d_{ur}$; (iii) all edges in the induced subgraph $G[R]$ are removed from E; (iv) all vertices in R except from r are removed from V.

Corollary 1 *Before contraction, the token graph $G = (V, E)$ constructed by generation and projection has $|V| - 2k$ vertices (from the $(k + 1)$-st to the $(n - k)$th) having the same node degree $2k$.*

Proof By Proposition 1, the graph $G = (V, E)$ can be constructed by projection first and then generation. Therefore this graph is a k-proximity graph, where the ith vertex has degree $2k$ for all $k < i \leq n - k$. □

Corollary 1 shows that the contraction step is essential to yielding proximity gows with range of different vertex degrees. This feature is important insofar as our aim is to look at word ranking functions based on vertex degrees in gows rather than word frequencies in documents.

2.2 Dependency

A dependency tree is by definition a tree graph over the sentence tokens. The **generation** of dependency trees from sentences is carried out by either Probabilistic Context-Free Grammar (PCFG) parsers [9] or appropriately trained neural networks [5].

Projection and **contraction** are the same as for proximity-based gows. We note that the projection step on dependency trees has a weak impact on connectivity, since most of the important tokens are naturally set at nodes closer to the root than non-important ones.

2.3 Constituency

A constituency tree is a tree graph over sentence tokens as well as syntax tags. In this sense, constituency trees can be seen as "liftings" from dependency trees. To a given constituency tree, one can retrieve the corresponding dependency tree[1] [7]. Vice-versa, there may be more than one constituency tree corresponding to a given dependency tree [18]. Existing algorithms aim at finding the smallest corresponding constituency tree.

The **generation** of constituency trees from sentences is carried out by either PCFG parsers (see https://nlp.stanford.edu/software/srparser.html) or appropriately trained neural networks (see https://pypi.org/project/benepar/).

Because constituency trees have more nodes than just tokens from the given sentence, a preliminary projection step is necessary to remove all of the non-token nodes. This is different from the projection step in proximity and dependency gows, because the impact on connectivity when removing grammatical tag nodes is considerable. We therefore defined a more connectivity-aware variant of projection: (i) for any pair (u, v) of leaf nodes (word tokens) in the constituency tree T of the sentence s, compute the shortest path $u \to v$ in T having length ℓ, and add the edge $\{u, v\}$ with weight $d_{uv} = \ell$ to the graph; (ii) remove all arcs adjacent to at least one non-leaf node; (iii) remove all non-leaf nodes. This preliminary projection step transforms the constituency tree into a graph on the word tokens from the sentence s.

We note that the most efficient algorithm for computing shortest paths in trees is by means of the Lowest Common Ancestor (LCA) of the origin and destination nodes. This yields a linear-time shortest path algorithm.

Projection and **contraction** are the same as for proximity-based gows.

3 Computational Experiments

Our benchmark aims at establishing advantages and disadvantages of different types of gows in keyword extraction tasks. We consider two corpora: a literary one, and a technical one. We extract keywords from documents in these corpora using the following rank functions: TFIDF and BM25 using term frequency, and TWIDF on k-proximity, constituency tree, and dependency tree based gows (see Sect. 1.1).

Our code is written in Python 3.10. For dependency and constituency syntax trees we made use of `spaCy` 3.4.4 [5] and `benepar` 0.2.0 [8]. Graphs were encoded and handled in NetworkX [4] 2.8.6. Experiments were obtained on an Apple M1 Max CPU with 64GB RAM and MacOS 12.6.3. See http://www.github.com/leoliberti/syntaxGraphOfWords to access the code and the corpora.

3.1 The Literary Dataset

The literary corpus contains 18 short documents extracted from various literary work, each consisting of a single paragraph. The lexical and grammatical quality of these

[1] See https://github.com/wenkokke/dep2con.

Table 1 Comparative results on a set of paragraphs from various literary sources, from which we extracted the three highest-rank keywords with various methods. We report the number of keywords given by each method that is in the list of three keywords in the ground truth

Instance		TermFreq		Graphs-of-words			
Source	*kw*	TFIDF	BM25	1-*prox*	4-*prox*	*con*	*dep*
1177 b.C.	3	1	1	1	2	0	1
Crossings	3	1	1	1	1	0	0
The golden bough	3	1	1	0	0	0	0
Illuminating Eco	3	0	0	0	0	0	0
The island of the day before	3	1	1	1	2	1	1
The library of Babel	3	0	0	0	0	0	1
Media stories: Malvinas	3	1	1	1	1	1	0
Neverwhere	3	2	2	1	1	0	1
Nothing	3	0	0	0	0	0	0
Paine	3	0	0	0	0	0	0
Foucault's Pendulum	3	0	0	0	0	0	0
The perks of being a wallflower	3	1	1	1	1	0	0
Quantum computing since Democritus	3	0	0	0	0	0	0
Richard III	3	1	1	1	1	0	1
The seventh function of language	3	0	0	0	0	0	0
Walden	3	1	1	0	0	0	0
When the sleeper wakes	3	1	1	1	1	0	0
Wisdom	3	0	0	0	0	0	1
Total	54	***11***	***11***	*8*	*10*	*2*	*6*

excerpts is perfect. The ground truth is a set of three keywords per document. These keywords were established by the authors of this paper before obtaining the computational results (we admit nonetheless to a considerable risk of personal bias in our ground truth).

The keywords extracted automatically from the literary corpus are the 3 topmost ranking ones according to the values of term frequency and graph degree rank functions. In Table 1 we report the number of keywords guessed by the automatic methods that are part of the set of keywords in the ground truth.

We see from Table 1 that term frequency based ranking methods are better than gow-based methods. Amongst the latter, 4-proximity gows yield the best performance. We also note that the two term frequency based rankings have exactly the same performance.

3.2 The Technical Dataset

The technical corpus consists of 449 documents, each of which is a client question to technical support. The corresponding ground truth was collected by one of the authors of this paper (NB) in the course of his work at OneTeam. The questions are "as asked", with the normal amount of lexical quirks and ungrammatical phrases. These documents are short (8.6 words on average). We therefore restricted k-proximity to $k = 1$, otherwise the central word in the sentence would have ended up having an abnormally high vertex degree in the k-proximity gow. The average number of keywords per document in the ground truth is 2.4, but the maximum is 5: we therefore allowed the extraction of up to 5 keywords (the gows often had fewer than five vertices, however).

In Table 2 we present comparative statistical distributions on the success scores of each method on documents with a certain number of ground truth keywords. Each entry has the format $x@y$ to mean that a given method was able to find y correct keywords x times, when ranking the *docs* documents having |GT| keywords in their ground truth. The total $9 + 238 + 143 + 36 + 3 = 428$ falls short of the total of 449 documents since 21 documents had no keywords. Moreover, the marginal sums do not match *docs* because we did not print the number of times methods found zero correct keywords (it suffices to subtract the marginal sums from *docs*).

Table 2 Comparative statistics on the technical corpus. Under "Input" we report the number (*docs*) of documents having |GT| keywords in the ground truth. Each data entry $x@y$ in row (|GT|, *docs*) and method-indexed column means that the corresponding method found y out of |GT| ground truth keywords in x documents

Input		Ranking method				
|GT|	*docs*	TermFreq		Graphs-of-words		
		TFIDF	BM25	1-*proximity*	*Constituency*	*Dependency*
1	9	6@1	6@1	6@1	6@1	6@1
2	238	111@1	111@1	**115@1**	**115@1**	113@1
		14@2	14@2	14@2	14@2	**15@2**
3	143	41@1	41@1	41@1	**43@1**	42@1
		76@2	76@2	**77@2**	76@2	**77@2**
		1@3	1@3	1@3	1@3	1@3
4	36	17@1	17@1	17@1	17@1	17@1
		6@2	6@2	6@2	6@2	6@2
		0@3	0@3	0@3	**1@3**	0@3
		1@4	**1@4**	**1@4**	0@4	**1@4**
5	3	3@1	3@1	3@1	3@1	3@1
Total	*428*	*276*	*276*	*281*	***282***	*281*

In this experiment we find that gows are more effective at keyword extraction than term frequency. Constituency tree based gows are marginally better than other gows. We also note, again, that the two term frequency based methods attain equal performance levels.

4 Conclusion

We looked at graphs-of-words constructed using syntax trees, and their performance in extracting keywords from text. There is no clear dominance of term frequency versus graph-of-words rankinds. Graph-of-words scored better with short ungrammatical sentences, term frequency in literary texts. In the future, we may apply this technique to structures such as "knowledge graphs", which can be obtained by mapping the words in the text into structured knowledge sources.

References

1. Chomsky, N.: Syntactic Structures. Mouton, The Hague (1956)
2. Eco, U.: Semiotics and the Philosophy of Language. Indiana University Press, Bloomington, IN (1984)
3. Franciscus, N., Ren, X., Stantic, B.: Dependency graph for short text extraction and summarization. J. Inf. Telecommun. **3**(4), 413–429 (2019)
4. Hagberg, A., Schult, D., Swart, P.: Exploring network structure, dynamics, and function using `NetworkX`. In: Varoquaux, G., Vaught, T., Millman, J. (eds.) Proceedings of the 7th Python in Science Conference (SciPy2008), pp. 11–15. Pasadena, CA (2008)
5. Honnibal, M., Montani, I.: Industrial-Strength Natural Language Processing in Python. spaCy (2023)
6. Hopcroft, J., Ullman, J.: Introduction to Automata Theory, Languages, and Computation. Addison-Wesley, Reading, MA (1979)
7. Johansson, R., Nugues, P.: Extended Constituent-to-Dependency conversion for English. In: Proceedings of the 16th Nordic Conference of Computational Linguistics, NODALIDA, pp. 105–112 (2007)
8. Kitaev, N., Klein, D.: Constituency parsing with a self-attentive encoder. In: Proceedings of the 56th Annual Meeting of the Association for Computational Linguistics, vol. 1 (Long Papers), pp. 2676–2686. ACL (2018)
9. Klein, D., Manning, C.: Accurate unlexicalized parsing. In: Proceedings of the 41st Meeting of the Association for Computational Linguistics, pp. 423–430 (2003)
10. Levine, R., Mason, T., Brown, D.: Lex and Yacc, 2nd edn. O'Reilly, Cambridge (1995)
11. Meyer, F.: English Corpus Linguistics. CUP, Cambridge (2004)
12. Mihalcea, R.: Graph-based ranking algorithms for sentence extraction, applied to text summarization. In: Proceedings of the 42nd Annual Meeting of the Association for Computational Lingusitics, volume Companion Volume of *ACL* (2004)
13. Mihalcea, R., Tarau, P.: Textrank: bringing order into text. In: Proceedings of the Conference on Empirical Methods in Natural Language Processing, pp. 404–411 (2004)
14. Pérez-Agüera, J., Arroyo, J., Greenberg, J., Perez Iglesias, J., Fresno, V.: Using BM25F for semantic search. In: Proceedings of 3rd International Semantic Search Workshop, pp. 1–8 (2010)

15. Roelleke, T., Wang, J.: TF-IDF uncovered: a study of theories and probabilities. In: Proceedings of the 31st Annual International ACM SIGIR Conference on Research and Development in Information Retrieval, pp. 435–442 (2008)
16. Rousseau, F.: Graph-of-words: Mining and retrieving text with networks of features. Ph.D. thesis, LIX, Ecole Polytechnique, France (2015)
17. Rousseau, F., Vazirgiannis, M.: Graph-of-word and TW-IDF: new approach to ad hoc IR. In: Proceedings of CIKM, New York, 2013. ACM
18. Xia, F., Palmer, M.: Converting dependency structures to phrase structures. In: Allan, J. (ed.) Proceedings of the First International Conference on Human Language Technology Research, HLT, San Francisco, 2001. Morgan Kaufman

On the Optimality Gap of Full Airport Slot Assignments: Capacity-Limited Packing with Pareto Optimality Constraints

Andreas Brieden, Peter Gritzmann, and Michael Ritter

Abstract We study a combinatorial packing problem with Pareto optimality constraints which arises naturally in the aviation industry. In fact, it has been observed that the current practice of assigning takeoff and landing rights at major airports may result in a significant gap between full and maximal flight schedules in practice. We analyze the specific packing problem theoretically and, particularly, study the occurring optimality gap under the prevailing regulatory regimes at the major airports in Germany. Finally, we report on the findings of a computational study based on real-world flight requests for one highly congested German airport.

1 Introduction

The allocation of takeoff and landing rights at congested airports is governed by constraints which restrict the number of arrivals, departures and movements in time intervals of specified length. More precisely, under the *International Air Transport Association*'s (IATA) slot system, the aircraft movements at a so-called coordinated airport are organized in the form of slots, which designate the right of an airline to execute a landing or a takeoff at a specified time. The standard scheduling interval has a length of 10 min, and can accommodate a specified number of arrivals, departures or, generally, movements. Further restrictions apply for other time periods, e.g., for intervals of 30 or 60 min length. These intervals are usually "rolling", i.e., the specified bounds need to hold for each interval of that length starting at any of the

A. Brieden
Universität der Bundeswehr München, D-85579 Neubiberg, Germany
e-mail: andreas.brieden@unibw.de

P. Gritzmann · M. Ritter (✉)
Department of Mathematics, TUM School of Computation, Information and Technology, Technical University of Munich, Munich, Germany
e-mail: michael.ritter@tum.de

P. Gritzmann
e-mail: gritzmann@tum.de

A. Brieden et al. (eds.), *Graphs and Combinatorial Optimization: from Theory to Applications*, AIRO Springer Series 13,
https://doi.org/10.1007/978-3-031-46826-1_15

Table 1 Reference value system applied at Frankfurt Airport. The R60M value of 106 includes 2 movements reserved for ad-hoc operations

Name	Length (min)	Operation	Type of constraint	Upper bound
R10A	10	Arrival	Fixed	13
R10D	10	Departure	Fixed	13
R10M	10	Movement	Fixed	20
R30A	30	Arrival	Rolling	33
R30D	30	Departure	Rolling	33
R30M	30	Movement	Rolling	57
R60A	60	Arrival	Rolling	60
R60D	60	Departure	Rolling	60
R60M	60	Movement	Rolling	106

10 min intervals, i.e., at the full hour, 10 min after the hour, 20 min after the hour etc. Table 1 depicts the current *slot regime* or *reference value system* at Frankfurt Airport as an example. Such reference value systems are designed to balance the number of overall flights, arrivals and departures for safety, security, service, environmental and other concerns.

At the beginning of the planning for a new season, all airlines submit either single *flight requests* or *flight series requests*. The former concern a single arrival or departure at a specified time or an arrival-departure pair with specified landing and takeoff times. The latter demand a series of movements at given times and days of the season. For instance, one such flight series request could ask for slots for the following operation:

> Arrival at 8:00, departure at 8:40 every Monday, Wednesday and Friday, starting June 1, ending September 30.

In practice, the airlines usually allow some flexibility in terms of the arrival and departure times, specify minimal ground time etc.; exact parameters are given with the request in the form specified by the responsible authorities.

The coordination of all flight requests for all coordinated airports in Germany is conducted by the "Flughafenkoordination Deutschland", a federally owned company; similar institutions exist throughout the European Union and in other countries. The final allocation which takes the inter-airport dependencies into account is then negotiated at an international flight scheduling conference.

Under the so-called "use-it-or-lose-it" provision, airlines receive a *grandfather right* for the upcoming summer or winter season for each flight request which was already assigned in the previous summer or winter season, respectively, and in operation for at least 80% of the time. The slots with grandfather rights must be assigned first before the remaining slot pool is allocated (with some additional constraints concerning new entrants).

There are various reasonable objectives for the allocation including that of welfare optimization (which, however, is currently hardly used in practice). Here, we will mainly focus on the total number of flight movements that can be allocated and refer to the underlying optimization problem as MAXMOV. See [2] for background information and detailed pointers to the literature, [2, 8, 9] for integer programming models and [2] for a study of auction based allocation.

As it has been observed in practice, the "interplay" between the slot regime and grandfather rights may lead to quite significant "blocking" which prevents any additional request from being assigned while the schedule contains a substantially lower number of movements than expected.

The paper is organized as follows. In Sect. 2 we will provide a combinatorial packing model with Pareto optimality constraints, and state our main results. Section 3 addresses computational complexity issues. In Sect. 4 we study the blocking effect for each fully coordinated airport in Germany based on its current individual reference value system. We model the problem of minimizing the number of flight movements under Pareto optimality as an integer programming problem and compute bounds for the worst-case optimality gap which is caused by flight requests, i.e., of single movements or arrival-departure pairs alone. We complement these results by reporting on the findings of a computational study based on real-world flight series requests for a major German airport in the presence of grandfather rights. Section 5 closes with some final remarks.

2 Notation, Preliminaries, and Main Results

Mathematically, slot assignment is a specific packing problem, and grandfather rights can be viewed as additional constraints which fix part of the packing. We will now introduce a combinatorial packing model with Pareto optimality constraints. It is slightly more general than what is required for our aviation problem. Its ingredients will still be interpreted within the realm of slot assignments.

For $n \in \mathbb{N}$, let $[n] = \{1, \ldots, n\}$, let $\mathbb{1} = (1, \ldots, 1)^T \in \mathbb{R}^n$, and, for $j \in [n]$, let u_j denote the jth standard unit vector in $\mathbb{R}^n$. Now, for $m, n \in \mathbb{N}$, let

$$A = (\alpha_{i,j})_{i\in[m],j\in[n]} = \begin{pmatrix} a_1^T \\ \vdots \\ a_m^T \end{pmatrix} \in \mathbb{Z}^{m\times n}, \quad b = \begin{pmatrix} \beta_1 \\ \vdots \\ \beta_m \end{pmatrix} \in \mathbb{Z}^m, \quad c = \begin{pmatrix} \gamma_1 \\ \vdots \\ \gamma_n \end{pmatrix} \in \mathbb{N}_0^n.$$

As usual, in the following, inequalities are meant component-wise.

In the slot assignment problem, the columns of the matrix A correspond to the potential time slots for each flight request while the rows encode the constraints imposed by the reference value system. The right hand side b contains the individual reference values, and c is the objective function vector, e.g., $c = \mathbb{1}$ when the number of allocated flight assignments has to be maximized.

In the following, the triple (A, b, c) will be regarded as a *packing task*, and the collection of all such tasks will constitute our *packing problem*. We will generally suppose that the polyhedron

$$P = \{x \in \mathbb{R}^n : Ax \leq b \wedge 0 \leq x \leq \mathbb{1}\}$$

contains at least one integer point, i.e., $P \cap \mathbb{Z}^n \neq \emptyset$. This is, of course, trivially satisfied when $b \geq 0$ since the empty packing 0 is always feasible then. Each vector $g \in P \cap \mathbb{Z}^n$ is referred to as a *packing* while the objective function value $c^T g$ is called its *density*. We speak of (A, b, c) as a *combinatorial packing task* if $A \in \{0, 1\}^{m \times n}$, $b \in \mathbb{N}^m$, and $c = \mathbb{1}$. Note that MAXMOV, i.e., the slot assignment problem with the goal to allocate as many movements as possible, is a combinatorial packing problem.

In the present paper we are particularly interested in the effect that grandfather rights will have on the maximal density which can be achieved. More generally, we study the worst-case gap between maximal and "full" packings. As usual, "maximal" refers to the global maximum $\max\{c^T g : g \in P \cap \mathbb{Z}^n\}$ of the objective function over all packings. On the other hand, "full" relates to a local optimum, i.e. a packing which does not allow the addition of any further object without violating at least one of the constraints.

Hence we say that a point $g \in P \cap \mathbb{Z}^n$ is *full*, if it is *Pareto-optimal* with respect to all coordinate directions, i.e.,

$$\hat{g} \in P \cap \mathbb{Z}^n \wedge \forall_{i \in [n]}\, u_i^T g \leq u_i^T \hat{g} \implies g = \hat{g}.$$

Let G_P denote the corresponding *Pareto front*, i.e., the set of all integer points in P which are full. Then we consider the following decision problem:

Problem 1 MINPARETOMAX

Input: $\gamma \in \mathbb{N}$, $A \in \mathbb{Z}^{m \times n}$, $b \in \mathbb{Z}^m$, $c \in \mathbb{N}_0^n$ such that $P \cap \mathbb{Z}^n \neq \emptyset$, where $P = \{x \in \mathbb{R}^n : Ax \leq b \wedge 0 \leq x \leq \mathbb{1}\}$.
Task: Decide whether there exist $g^* \in P \cap \mathbb{Z}^n$ and $g_* \in G_P$ such that $c^T g^* - c^T g_* \geq \gamma$?

Of course, MINPARETOMAX is the decision version of computing

$$\max\{c^T x : x \in P \cap \mathbb{Z}^n\} - \min\{c^T x : x \in G_P\}$$

which asks for the *optimality gap*, i.e., the difference in the objective function values of a maximal and a minimal full packing.

In slot allocation we are interested in the effect of a priority assignment for requests which are endowed with grandfather rights. The optimality gap is then an upper bound for the consequences of current assignment regimes for the degree of exhaustion of airport capacity.

Gaps between global and certain local optima are well studied for packings of objects of different sizes or weights, with a particular view on the behavior of specific approximation algorithms. The area of *bin packing*, for instance, provides a large class of examples; see, e.g., [4]. The existence and extent of such effects is, however, less obvious for packings of objects of the same weight and same gain. To the best of our knowledge, optimality gaps which are caused by the interplay of grandfather rights with specific reference value systems have not been studied systematically before.

As it turns out, even quite restricted versions of MINPARETOMAX are $\mathbb{NP}$-complete. For background material on computational complexity see [5].

Theorem 1 MINPARETOMAX *is* $\mathbb{NP}$*-complete. The* $\mathbb{NP}$*-completeness persists even if all instances are restricted to those for which* $\gamma = 1$*, one integer solution is explicitly known, all integer solutions have the same cardinality, all entries of the matrices and right hand sides are in* $\{0, \pm 1\}$*,* $c \in \{0, 1\}^n$*, and* $(n, m) = (p^2, 8p - 4)$ *with* $p \in \mathbb{N}$.

The special case of the optimization version of MINPARETOMAX where all instances are combinatorial packing tasks will be referred to as COMBMINPARETOMAX. We show that COMBMINPARETOMAX is hard to approximate, even when the right hand side is restricted to the all-ones vector.

Theorem 2 COMBMINPARETOMAX *is* $\mathbb{APX}$*-hard even if all instances are restricted to those for which* $b = \mathbb{1}$.

In the context of slot allocation, Theorem 2 can be viewed as an indication that (unless $\mathbb{P} = \mathbb{NP}$) there will never be a practically efficient deterministic algorithm for computing the optimality gap in general. Note, however, that for any fixed reference value system the number of different slot assignments within a season is a polynomial in the number of flight requests whose degree is a function of the reference values. Given the figures of Table 1 it is clear that the gigantic degrees of these polynomials for real-world instances render any enumerative approach hopeless in practice.

In Sect. 4 we model the problem of minimizing the number of flight movements under Pareto optimality as an integer programming problem and compute bounds for the worst-case optimality gap. Our computations indicate that (even without the presence of longer flight series) the potential blocking effect for each fully coordinated airport in Germany based on its current individual reference value system is at least 8.9% (with one notable exception) and may be as large as 15.9%, depending on the slot regime. For details see Tables 3 and 4. These results are complemented by the findings of a computational study for a major German airport based on the real-world flight series requests.

3 Computational Complexity

Here we will only briefly indicate which problems are used for the reduction but refer to the full version [3] of this paper for the details. First note that, while integer linear programming is $\mathbb{NP}$-complete, it is not clear a priori whether the hardness persists in our special situation. Accordingly, we use reductions from rather different problems in order to establish our complexity results.

Obviously, MinParetoMax is in $\mathbb{NP}$. Its $\mathbb{NP}$-hardness can be proved by means of a reduction from a specific uniqueness problem from *discrete tomography*, see [1].

In order to show that CombMinParetoMax is $\mathbb{APX}$-hard even if all instances are restricted to those for which $b = \mathbb{1}$ we can use a reduction from the following set packing problem.

Problem 2 Maximum 3- Set Packing

Input: A finite set C with $|C| \geq 4$, and a set $\mathcal{C} \neq \emptyset$ of nonempty subsets of $C$of cardinality 3.
Task: Find the maximum cardinality of all packings in $\mathcal{C}$, i.e., of all subsets $\mathcal{S}$ of $\mathcal{C}$ whose elements are disjoint.

As [7] showed, Maximum 3- Set Packing is $\mathbb{APX}$-complete, even when restricted to instances where, for some $q \in \mathbb{N}$ with $q \geq 3$, no element of C occurs in more than q of the subsets in $\mathcal{C}$. Let us remark that, by [6], Maximum k- Set Packing can be approximated within a factor $k/2 + \varepsilon$ for any $\varepsilon > 0$. However, this does not seem to imply particularly strong approximation results for CombMinParetoMax.

4 Gap Potential at the German Level 3 Airports

In the following we will analyze the potential optimality gaps at all German Level 3 airports, i.e., airports operating at the limit of their capacity which are therefore subject to coordination (Berlin, Düsseldorf, Frankfurt, Hamburg, Hannover, München and Stuttgart). The current reference value systems (summer season 2023) for these airports are given in Table 2.

The table is slightly simplified: the figures apply only for the core times (in reality, the limits sometimes vary slightly according to the time of day), slots for ad-hoc operations are treated as regular slots, and specific values for flights on a North Atlantic route are omitted; see https://fluko.org for the precise information. Generally, the reference values are "rolling", i.e., they apply for each interval of the specified length beginning every 10 min (except for Düsseldorf airport).

First, we model the optimization version of MinParetoMax as an integer programming problem. So, let (A, b, c) be a given instance. Using the obvious decou-

Table 2 Reference value system for the summer season 2023, slightly simplified

	10 min			20 min			30 min			60 min		
	A	D	M	A	D	M	A	D	M	A	D	M
Berlin	12	12	15	–	–	–	32	32	45	50	50	78
Düsseldorf	8	7	12	–	13	–	–	–	27	33	36	47
Frankfurt	13	13	20	–	–	–	33	33	57	60	60	106
Hamburg	7	7	9	–	–	–	18	18	25	31	31	48
Hannover	6	6	8	–	–	–	–	–	–	30	34	40
München	12	12	15	–	–	–	–	–	–	58	58	90
Stuttgart	7	7	9	–	–	–	–	–	–	35	35	48

pling into the maximization $\max\{c^T x : x \in P \cap \mathbb{Z}^n\}$ and the minimization problem $\min\{c^T x : x \in G_P\}$ it suffices to consider the latter.

Note that $y \in G_P$, if and only if $y \in P \cap \mathbb{Z}^n$, but $y + u_j \notin P$ for any $j \in [n]$, i.e., there exists an index $i \in [m]$ such that $A(y + u_j) \not\leq b$ or $y + u_j \not\leq \mathbb{1}$. Hence, $y \in G_P$, if and only if, $y \in P \cap \mathbb{Z}^n$, and for every $j \in [n]$ with $u_j^T y = 0$ there exists an index $i \in [m]$ such that

$$a_i^T (y + u_j) = a_i^T y + \alpha_{i,j} \geq \beta_i + 1,$$

where $a_1^T, \ldots, a_m^T$ denote again the rows of A. Setting

$$\mu_i = \alpha_{i,j} - (\beta_i + 1) + \min\{a_i^T x : x \in P\},$$

the condition can be written as

$$\begin{aligned} & a_i^T y + \mu_i \delta_{i,j} \geq \beta_i + 1 - \alpha_{i,j} + \mu_i && (i \in [m],\ j \in [n]) \\ & \delta_{i,j} \leq 1 - u_j^T y, \qquad \delta_{i,j} \in \{0, 1\} && (i \in [m],\ j \in [n]), \\ & \sum_{i=1}^{m} \delta_{i,j} \geq 1 && (j \in [n]). \end{aligned}$$

Note that the constraints $\delta_{i,j} \leq 1 - u_j^T y$ imply that $\delta_{i,j} = 0$ if the jth component of y is already 1.

Hence, using linear programming to compute $\mu_1, \ldots, \mu_m$, and setting $\tau_{i,j} := \beta_i + 1 - a_{i,j} + \mu_i$ for $i \in [m]$, $j \in [n]$, the minimization problem $\min\{c^T y : y \in G_P\}$ can be formulated as

$$\begin{aligned} \min\Big\{c^T y : Ay \leq b \wedge y \in \{0, 1\}^n \wedge a_i^T y + \mu_i \delta_{i,j} \geq \tau_{i,j} \wedge \delta_{i,j} \leq 1 - u_j^T y \\ \wedge \sum_{i=1}^{m} \delta_{i,j} \geq 1 - u_j^T y \wedge \delta_{i,j} \in \{0, 1\} \quad (i \in [m],\ j \in [n])\Big\}. \end{aligned}$$

We used essentially this formulation to compute bounds on the optimality gap for all level 3 airports in Germany based on their slot regime according to Table 2 and the following assumptions:

- A complete day consists of 18 h (usually 5:00 a.m. to 10:59 p.m. but times may vary from airport to airport due to nighttime flight restrictions).
- For each time during the hours of operation an arbitrary number of flight requests for arrivals and departures is available (i.e., we are just concerned with the number of arrivals and departures, not with actual flight or flight series requests comprising additional constraints).
- A flight request involves either a single movement (arrival or departure) or a tightly coupled arrival-departure pair where the departure is exactly 60 min after the corresponding arrival (on the runway).

As, due to the Pareto constraints, the integer linear programs are computationally too challenging for obtaining provably optimal solutions for the full range of operating hours within reasonable time we resorted to the following method for obtaining lower bounds on the optimality gap:

- First, the maximum gap and corresponding maximal and minimal full solutions were computed for all time periods between 1 and 4 h for each airport. The time windows for the reference value systems were "wrapped around" to make sure that these "patterns" could be concatenated to obtain a feasible solution for longer time periods.
- A feasible solution for a period of exactly 18 h was subsequently obtained by concatenating one of these solutions a sufficient number of times and then adding additional 10-minute intervals to the beginning and the end of the time period such that at least one full hour was still available both at the beginning and at the end of that solution.
- Flight movements (arrivals and departures) for the intervals added at the beginning and at the end were computed (using a straightforward modification of the MINPARETOMAX integer linear program) such that the overall gap of the solution over the total of 18 h was maximized. No coupling between arrivals and departures was enforced for these additional flights.

The solutions obtained in this way naturally provide lower bounds for the possible gap, and we selected those which provided the largest gap. The numbers of flight movements for the so obtained full solutions, maximum solutions and the resulting bound for the optimality gap are depicted in Tables 3 and 4. Of course, the actual optimum might have an even larger gap.

As Table 3 shows the obtained bound for the optimality gap for Düsseldorf Airport is 0. In fact, a case analysis shows that, under our assumptions on the flight requests, the optimality gap at Düsseldorf is actually 0. The main difference of the slot regime at Düsseldorf is that the bounds on the arrivals, departures and movements are not required for *rolling* intervals while the reference values at the other airports are rolling. Of course, coupling effects inflicted by complex flight series requests for

Table 3 A lower bound on the worst-case optimality gap for the valid reference value systems under the conditions specified above for the case of no arrival-departure coupling

	min. full	Maximum	Gap	Ratio (% of max.)
Berlin	1202	1404	202	14.4
Düsseldorf	846	846	0	0
Frankfurt	1738	1908	170	8.9
Hamburg	745	864	119	13.8
Hannover	624	720	96	73.3
München	1362	1620	258	15.9
Stuttgart	786	864	78	9.0

Table 4 A lower bound on the worst-case optimality gap for the valid reference value systems under the conditions specified above for the case where arrivals and departures are tightly coupled with exactly 60 min between an arrival and its corresponding departure (except for the extra flights added at the boundary)

	min. full	Maximum	Gap	Ratio (% of max.)
Berlin	1082	1404	322	22.9
Düsseldorf	743	846	103	12.2
Frankfurt	1665	1906	241	12.6
Hamburg	654	864	210	24.3
Hannover	538	720	182	25.3
München	1181	1616	435	26.9
Stuttgart	654	210	210	24.3

longer time periods may still result in significant blocking, but the reference value system itself is "robust" at least against the most simple manipulation attempts.

To facilitate the understanding of the underlying patterns, let us take a look at an explicit solution for Frankfurt airport for the case of tightly linked arrivals and departure pairs with exactly 60 min between an arrival and the corresponding departure. As we will see, even in this extremely restricted situation the optimality gap is quite significant. The solution is based on a pattern for 12 consecutive 10-minute intervals, thus it is sufficient to look at four hours. We have selected the time period from 6:00 a.m. to 10:00 a.m. as that is usually a time of high traffic at Frankfurt airport and thus the actual flight movements might be somewhat close to the "prototypical" case we have computed.

In the slot assignment indicated in Table 5, in five out of six 10-minute intervals either arrivals or departures are at their maximum value of 13. The remaining movements are determined such that R30M is at maximum for the first three intervals, thus completely covering the first five 10-minute intervals. For the sixth such interval, the numbers are chosen such that R60A is full for the intervals 1 to 6, i.e., 06:00 a.m. to 06:50 a.m. and R60D is full for the intervals 6-12, i.e., 06:50 a.m. to 07:50 a.m.

Table 5 Part of the minimum full configuration computed for Frankfurt airport; arrivals and departures are tightly coupled with a difference of 60 min. The gray overlays indicate that the corresponding 10-minute intervals (rows) are "blocked" by an active rule, boldface indicates a rule at its limit. The values for R10 rules are equal to the number of arrivals, departures and movements and are thus not explicitly given. For the other rules, each component depicts the number of corresponding flights in the rolling interval beginning at the time indicated in its row

Time	A	D	M	R30A	R30D	R30M	R60A	R60D	R60M
06:00	**13**	7	**20**	**33**	24	**57**	**60**	36	96
06:10	**13**	4	17	**33**	24	**57**	54	42	96
06:20	7	**13**	**20**	**33**	24	**57**	45	51	96
06:30	**13**	7	**20**	27	12	39	51	45	96
06:40	**13**	4	17	21	18	39	45	51	96
06:50	1	1	2	12	27	39	36	**60**	96
07:00	7	**13**	**20**	24	**33**	**57**	36	**60**	96
07:10	4	**13**	17	24	**33**	**57**	42	54	96
07:20	**13**	7	**20**	24	**33**	**57**	51	45	96
07:30	7	**13**	**20**	12	27	39	45	51	96
07:40	4	**13**	17	18	21	39	51	45	96
07:50	1	1	2	27	12	39	**60**	36	96
08:00	**13**	7	**20**	**33**	24	**57**	**60**	36	96
08:10	**13**	4	17	**33**	24	**57**	54	42	96
08:20	7	**13**	**20**	**33**	24	**57**	45	51	96
08:30	**13**	7	**20**	27	12	39	51	45	96
08:40	**13**	4	17	21	18	39	45	51	96
08:50	1	1	2	12	27	39	36	**60**	96
09:00	7	**13**	**20**	24	**33**	**57**	36	**60**	96
09:10	4	**13**	17	24	**33**	**57**	42	54	96
09:20	**13**	7	**20**	24	**33**	**57**	51	45	96
09:30	7	**13**	**20**	12	27	39	45	51	96
09:40	4	**13**	17	18	21	39	51	45	96
09:50	1	1	2	27	12	39	**60**	36	96

This pattern of R60A/R60D alternates with one 10-minute interval overlap at 06:50 a.m., 07:50 a.m. etc. where R30M takes care of the intervals "in between". In effect, the rules are combined to obtain a full flight schedule with as few movements as possible.

Table 6, on the other hand, depicts a maximum configuration for the same time period. Note that all R60M rules are now binding, showing that the assignment is indeed maximal. Also R10M is satisfied with equality as much as possible.

When we extend the depicted schedules to the full time period from 6 a.m. to 11 p.m. as described above, we obtain two different full schedules which however,

Table 6 Part of a maximum configuration computed for Frankfurt airport; arrivals and departures are tightly coupled with a distance of 60 min. The gray overlays indicate that the corresponding slots (rows) are "blocked" by an active rule, boldface numbers indicate an active limit. The values for R10 rules are equal to the number of arrivals, departures and movements and are thus not explicitly given

Time	A	D	M	R30A	R30D	R30M	R60A	R60D	R60M
06:00	7	**13**	**20**	24	**33**	**57**	48	58	**106**
06:10	7	**13**	**20**	30	22	52	54	52	**106**
06:20	10	7	17	**33**	19	52	**60**	46	**106**
06:30	**13**	2	15	24	25	49	57	49	**106**
06:40	10	10	**20**	24	30	54	46	**60**	**106**
06:50	1	**13**	14	27	27	54	46	**60**	**106**
07:00	**13**	7	**20**	**33**	24	**57**	58	48	**106**
07:10	**13**	7	**20**	22	30	52	52	54	**106**
07:20	7	10	17	19	**33**	52	46	**60**	**106**
07:30	2	**13**	15	25	24	49	49	57	**106**
07:40	10	10	**20**	30	24	54	**60**	46	**106**
07:50	**13**	1	14	27	27	54	**60**	46	**106**
08:00	7	**13**	**20**	24	**33**	**57**	48	58	**106**
08:10	7	**13**	**20**	30	22	52	54	52	**106**
08:20	10	7	17	**33**	19	52	**60**	46	**106**
08:30	**13**	2	15	24	25	49	57	49	**106**
08:40	10	10	**20**	24	30	54	46	**60**	**106**
08:50	1	**13**	14	27	27	54	46	**60**	**106**
09:00	**13**	7	**20**	**33**	24	**57**	58	48	**106**
09:10	**13**	7	**20**	22	30	52	52	54	**106**
09:20	7	10	17	19	**33**	52	46	**60**	**106**
09:30	2	**13**	15	25	24	49	49	57	**106**
09:40	10	10	**20**	30	24	54	**60**	46	**106**
09:50	**13**	1	14	27	27	54	**60**	46	**106**

accommodate 1738 and 1908 flights, respectively. Even though not a single additional arrival or departure can be incorporated in the first schedule, it contains 241, i.e., 12.6% less flights than the second one. Similar examples provide the figures of Tables 3 and 4 for the other level 3 airports in Germany.

Let us point out that, for various reasons, the derived ratios should only be viewed as theoretical indication for potential blocking within the slot regimes. First, in practice, a flight can only be assigned if slots are requested by an airline. Second, in the process of coordination one attempts to solve MaxMov (or a variant of it) and hence tries to avoid "bad" solutions as far as possible. However, as we have seen, blocking already occurs when only requests for single movements are considered and may thus be introduced inadvertently into the scheduling process.

Even worse, in practice, the majority of requests is not for single movements or arrival-departure pairs but for flight series at specified times, days and weeks throughout the season. In effect, such series constitute a "wide range coupling" throughout the season, which may result in a significantly higher optimality gap. Further recall that grandfather rights can be viewed as fixing partial solutions which may drive even best assignments of the remaining slot pool to full schedules of smaller cardinality.

In order to assess how these conflicting effects act in practice, we studied optimality gaps empirically at a major German airport under realistic assumptions and, in particular, based on the real flight requests for a full season which included information on the connection of arrivals and subsequent departures, on minimum ground times requirements etc. Also, a tolerance of plus/minus 10 min for intercontinental flights and of plus 10 min for continental flights compared to the requested slot time was permitted.

In the first scenario we solved MAXMOV under the assumption that no grandfather rights had to be observed. In the second and third scenario, 40% and 50% of the requests were endowed with grandfather rights, respectively. All computations were carried out for a full season (which is roughly six months long). Due to the grandfather rights, the maximum number of movements which could be scheduled in the second scenario was by about 4000 less than the number in scenario one. The presence of grandfather rights for 50% of the requests led to a reduction of about 6000 movements in the season.

5 Final Remarks

As we have seen, even a simple regime of flight requests carries the potential for significant optimality gaps. In the absence of grandfather rights suboptimal schedules can be avoided by using state-of-the-art integer programming algorithms which are capable of computing optimal schedules; see [2, 9].

In conjunction with the slot regime grandfather rights can act as additional constraints or can be viewed as a partial result of some scheduling heuristic and can hence lead to significant gaps between full and maximal schedules. It might therefore be advisable to study the effect of weakening grandfather rights to some extent. The right to operate at times close to but not quite exactly at the time of the previous usage will typically allow more flexibility in flight scheduling and thus lead to schedules with additional flights if needed. The effect of such changes will, of course, depend on the slot regime, the flight series requests, the amount and location of previous grandfather rights etc. and may hence vary from airport to airport.

References

1. Alpers, A., Gritzmann, P.: On the reconstruction of static and dynamic discrete structures. In: R. Ramlau, O. Scherzer (eds.) The Radon Transform: The First 100 Years and Beyond, pp. 297–342. De Gruyter, Berlin (2019). https://doi.org/10.1515/9783110560855-013
2. Bichler, M., Gritzmann, P., Karaenke, P., Ritter, M.: On airport time slot auctions: a market design complying with the IATA scheduling guidelines. Transp. Sci. **57**(1), 27–51 (2022). https://doi.org/10.1016/j.tcs.2008.06.014
3. Brieden, A., Gritzmann, P., Ritter, M.: On capacity-limited packing with pareto optimality constraints, and its application to airport slot assignments. Technische Universitä München, Technical Report (2023)
4. Coffman Jr., E., Csirik, J., Galambos, G., Martello, S., Vigo, D.: Bin packing approximation algorithms: Survey and classification. In: P. Pardalos, D. Du, R. Graham (eds.) Handbook of Combinatorial Optimization, pp. 455–531. Springer, New York, NY (2013). https://doi.org/10.1007/978-1-4419-7997-1_35
5. Garey, M., Johnson, D.: Computers and intractability. A guide to the theory of NP-completeness. A Series of Books in the Mathematical Sciences. W. H. Freeman and Company, San Francisco (1979)
6. Hurkens, C.A.J., Schrijver, A.: On the size of systems of sets every t of which have an SDR, with an application to the worst-case ratio of heuristics for packing problems. SIAM J. Disc. Math. **2**, 68–72 (1989)
7. Kann, V.: Maximum bounded 3-dimensional matching is MAX SNP-complete. Inform. Process. Lett. **37**, 27–35 (1991)
8. Ribeiro, N., Jacquillat, A., Antunes, A., Odoni, A., Pita, J.: An optimization approach for airport slot allocation under IATA guidelines. Transp. Res. B **112**, 132–156 (2018)
9. Ritter, M.: Packing under Balancing Constraints. Applications in Semiconductor Design and Flight Scheduling. Ph.D. thesis, Technical University of Munich (2008)

Author Index

A. Brieden et al. (eds.), *Graphs and Combinatorial Optimization: from Theory to Applications*, AIRO Springer Series 13,
https://doi.org/10.1007/978-3-031-46826-1